Front

The Universe

is

Alive and Well

The Organism

Katya Walter, Ph.D.

**The author thanks the
Institute for Neuroscience and Consciousness Studies
of Austin, Texas
for its helpful support and encouragement**

THIS BOOK IS
VOLUME 4, FOURTH EDITION
IN THE
TOUCHING GOD'S TOE SERIES

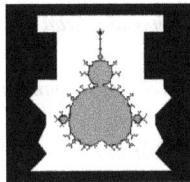

To visit THE DOUBLE BUBBLE UNIVERSE, GO TO......
https://www.katyawalter.com

To visit KATYA WALTER'S YOUTUBE CHANNEL, GO TO...
KatyaWalterYouTube

Kairos Center Publications
Box 142086
Austin, Texas 78714
kairospublications@gmail.com

The Universe is Alive and Well: The Organism by Katya Walter, Ph.D.
Volume 4 in the *Touching God's TOE series*, 4th edition
Copyright © 2004 by Kairos Center; 4th edition by Kairos Center
Editor: Jennie Rosenblum   Art by Adele Aldridge, Adrian Frye, & Katya Walter
                              - or from Wiki Commons or Creative Commons
Paperback 4th edition published 2019   ISBN 978-1-884178-53-5
Electronic 4th edition published 2019   ISBN 978-1-884178-78-8

Library of Congress Cataloging-in-Publication Data

Walter, Katya - *The Universe is Alive and Well: The Organism*
Includes table of contents, appendix, bibliography, & illustrations
1. Physics—gravity, cosmology, strings, spacetime, dimensions, fractal topology
2. Religion—Touching God's TOE, spirit & science, religions, divine love
3. Gravity—gravitation, unification, forces, emergent properties
4. Chaos Theory—Lorenz attractor, fractals, chaos patterning, complexity
5. Philosophy—Plato, Taoism, Chinese thought
6. Mathematics—nonlinear, analinear, fractals, analog & linear number
7. Mysticism—mystic love, remote viewing, dreams, I Ching, synchronicity systems
8. Title: *The Universe is Alive and Well: The Organism*

Katya Walter's Books
Chaosforschung (in German) - *1992 - Diederichs Verlag*
Dream Mail: Secret Letters for your Soul - 1995 - Kairos Center
Tao of Chaos: Merging East and West - 1994 - Kairos Center - This original book was split and augmented to become Volumes 2 and 3 of the *Touching God's TOE series*, first published in 2004, and updated in a 4th edition, as shown below.

Touching God's TOE series, 4th Edition

Vol. 1: Double Bubble Universe: The Paradigm	*2018*
Vol. 2: Co-Chaos Patterns: The Universal Fractal	*2019*
Vol. 3: Tao of Life: The Fractal Gift	*2019*
Vol. 4: The Universe Is Alive and Well: The Organism	*2020*
Vol. 5: Master Code Tree: The Expansions	*2022*
Vol. 6: Stone Soup Universe: The Hologram	*2023*
Vol. 7: The Particle Ark (projected)	
Vol. 8: Quantum Organics (projected)	

Table of Contents

> The sun, with all those planets revolving around it and dependent on it, can still ripen a bunch of grapes as if it had nothing else in the universe to do.
>
> *Galileo Galilei*

Introduction–What Is This Book?

From the Author: This fourth volume describes our Double Bubble universe as a living system composed of two giant mirror-twin bubbles that inflated during the Big Bang. We live in the familiar upper bubble that current science calls the whole universe. We exist far above the tiny quantum scale where particle-waves of matter and energy emerge at the lower limit of testable physics.

Our known huge bubble of 3D space is conjoined to a huge bubble of 3D time via a porous membrane interface whose myriad mobic pores provide constant communication between both bubbles. At this ultra-tiny mobic scale, the fine-grained dimensional fabric of space and time emerges with a built-in potential to hold matter and energy at a larger scale.

(In this TOE, space holds matter and time holds energy.)

How did this universe become doubled into conjoined mirror-twins? After making dimensionality, the cosmic egg split into a pair of conjoined mirror-twins when the Big Bang got around to manifesting matter and energy at the quantum scale. That late-stage quantum development is why the two bubbles are not identical twins, but instead, mirror-twins with reciprocal features.

Our known bubble got the features of contiguous 3D space, the arrow of time, the known pole of gravitation, and all the original antimatter. But that other bubble, the mirror-twin bubble unrecognized by current science? It got 3D time, the arrow of space, the "lost" pole of gravitation, and the original antimatter that science considers "disappeared" from this known bubble.

Moreover, that other bubble's arrow of space put such great pressure on the antimatter that it soon compress-converted into speedy tachyonic energy traveling in contiguous 3D time. That quick energy soon organized to power up a huge unified mind spread throughout that other bubble's 3D time. It gave the close, detailed, creative attention to this bubble that we call Mother Nature.

The two symbiotic bubbles constitute a single living system generated by a fractal master code that used a polarized pair of pairs—the space-time pair and the matter-energy pair—to generate and maintain the Double Bubble universe.

The master code is so fail-safe that it allowed the universe to foster its own evolution by iterating lesser coding variants at different scales. One such variant is the genetic code. It used a polarized pair of pairs—DNA's Thymine-Adenine pair and Cytosine-Guanine pair—to generate and maintain us.

The master code and genetic code both operate via 64 polarized, pair-bonded triplets. Fortunately, both of those fractal code variants can even be shorthanded by a third variant, the I Ching's ancient math that also uses 64 polarized, pair-bonded triplets.

This Double Bubble universe holds mind in many formats. Tiny, diverse minds are sprinkled across our own bubble's 3D space. A single, giant mind spreads across the other bubble's 3D time. Powered by super-swift tachyonic energy, its vast cloudbank of data is constantly updating at the membrane interface. That single great mind in the other bubble and our myriad tiny minds in this bubble work together to shape each life in the evolution of our universe.

Fractal resonances in the master code let the human unconscious, which is larger than ego, tap into its huge data bank in dreams, meditation, trance, psychic healing, precognition flashes, and other altered states. Its great mind is known to us by many names: Tao, Over-Soul, Akashic record, collective unconscious. Some call it God, but God is far greater than all the universes.

-:::-

From the Editor: This book is Volume 4 in the dazzling *Touching God's TOE* series, 4th edition. In this volume, Dr. Walter discusses our universe as a living system. She describes the master code that created its physical structure and the universal mind that vitalizes it. While offering scientific concepts, she also discusses philosophical parallels and dreams that inspired her to write this series.

This book has 18 chapters in 115 sections, with 79 listed images, graphics, and charts. It includes a *Series Summary*, *Bibliography*, and *Reviews*. The more science-based chapters have odd numbers (1, 3, 5, etc.). The more philosophical chapters have even numbers (2, 4, 6…). The ebook version has an interactive table of contents and 125 e-links that act as informative footnotes. Its text is searchable and receives electronic updates. It is hand-edited to hold color graphics that allow greater distinctions in images and charts. Consider getting both the print and ebook versions for a greater range of information and versatility.

Science and mysticism merge in this stunning new paradigm. China's ancient I Ching followed the flow of the Tao, universal mind. Western science investigates the genetic code that generates organic matter. This paradigm merges them to explain both mind and matter. Let Dr. Walter guide you through the patterns of chaos theory into mystic beauty! This book heals the 2,500-year split between body and mind. It embraces East and West to unite our planet. Called in Germany a *"philosopher queen of the global village"*… she *"merges left-brain accuracy with right-brain vision! Scientific truth speaks in a clear voice of wisdom."*- **Claus Claussen**

-:::-

Each book in this series has its own symbol. This volume's symbol suggests that our cities can become joyful urban islands rather than letting the sprawl of over-population pollute Earth's human and animal habitat.

Volume 4's symbol is written in the *Good Life* font by Qkila. Its eight letters form a "word"—*aguklezx*—in that font. Let's say it means *Good Luck!*

FOR AN OVERVIEW OF THE WHOLE SERIES, LOOK FOR THE SERIES SUMMARY AT THE BACK.

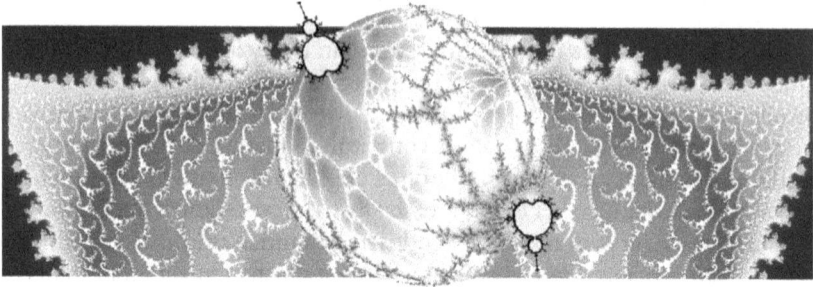

List of 20 Cosmological Questions

Earlier volumes in this series began with a discussion of 20 intriguing puzzles about the universe. This Volume 4 is mostly designed to show you how space and time are generated by polarized pulsing, and also how I Ching hexagrams can shorthand it, as well as describing some philosophical ideas. So in this volume, I'll list the 20 questions briefly to indicate the larger scope of the series.

In physics, many important questions still remain unanswered. This series examines 20 of these questions to see how they are answered in the Double Bubble TOE.

Question 1: What is our universe?

Question 2: What is the working shape of the universe?

Question 3: How did the universe begin?

Question 4: How deep in nature does fractal patterning go?

Question 5: Where did the original lost antimatter go?

Question 6: How did dimensions develop?

Question 7: What is gravitation?

Question 8: Why does the universe seem to expand constantly and ever-faster?

Question 9: Did the cosmic egg inflate in a hot Big Bang?

Question 10: Why do so many equations have important reciprocal solutions?

Question 11: Why is physics plagued with "impossible" infinities?

Question 12: What is electromagnetism? What is polarity?

Question 13: Why does light in a vacuum move at a constant speed? Not faster?

Question 14: Why won't Einstein's cosmological constant go away?

Question 15: Why can a particle-wave act like a particle OR a wave?

Question 16: How can two particles communicate faster than the speed of light?

Question 17: What is the neutrino?

Question 18: Is our universe designed to foster life?

Question 19: How will our universe end? Or will it?

Question 20: Is it possible to reconcile physics & philosophy in a TOE?

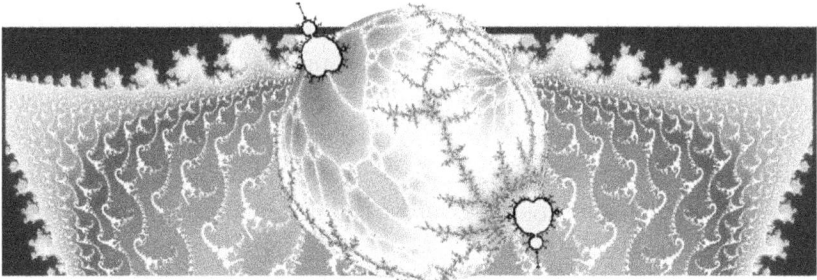

Chapter 1: The Universe Lives!

1. Fungus, Gaia, and Pachamama

This series says our universe is alive, and we are living inside it. Think of us as something like smart, self-aware bacteria inside the universal body, except relatively speaking, we are far smaller than bacteria inside a human body.

Is it hard to imagine that the universe itself is alive? What is the biggest living entity that springs to mind? An elephant? A whale? A sequoia? Very big things can be alive, even things so large that they seem outside our normal definition of a living organism. For example, central Utah holds a living being called Pando, a system of over 40,000 clonal aspen trees, with all of them connected by their roots. It is the heaviest known organism on Earth.

Or consider the largest (not heaviest) organism known to science. *Armillaria ostoyae* occupies about 2,384 acres in the Blue Mountains of Oregon. This fungus is at least 2,400 years old, but some estimate as much as 8,500 years old. Anne Casselman said in "Strange but True: The Largest Organism on Earth Is a Fungus," in *Scientific American*, "...the discovery of such huge fungi specimens rekindled the debate of what constitutes an individual organism."

Tom Volk, a biology professor at the University of Wisconsin–La Crosse, said that *Armillaria ostoyae* does meet the definition of a single living organism: "It's one set of genetically identical cells that are in communication with one another, that have a sort of common purpose or at least can coordinate themselves to do something." In other words, the fungus has genetically identical cells that communicate and coordinate for a shared common purpose.

In the 1970s, James Lovelock and Lynn Margulis introduced the scientific hypothesis that planet Earth is a single living entity because it exhibits the properties of communication, coordination, and common purpose. That hypothesis has been verified in many well-documented studies, and consequently, it was adopted in many life sciences, as well as in general systems study. By now, many scientists agree that every organism on Earth is closely integrated into the supersystem of a single, self-regulating holistic network.

Some earth system scientists call this holistic network *Gaia* in honor of a mythic Greek goddess who was Earth's divine personification. The modern Gaia hypothesis is based on much data that suggests this globe of Earth is a living system that includes its magnetic poles, dirt, oceans, atmosphere, plants, animals, fish, birds, and insects…with all of it communicating and responsive at some level, coordinating changes as needed to keep its bioenergetics going.

A widespread English folk name for this holistic network is "Mother Nature," who invisibly maintains life on this planet. It exhibits remarkably responsive interactions among the factors of geography, climate, plants, animals, insects, and so on, with all of it designed to keep the whole integrated system viable.

Even now, the Andes Quechua people call the Earth's living system *Pachamama,* a fertility goddess whose self-sufficient, creative power sustains life. Some earth system scientists, though, think names such as Gaia or Pachamama have undesired overtones of past religions—pantheism, deism, paganism, polytheism, shamanism—and they strive to avoid such connotations.

This TOE declares that our universe itself is alive, evolving, and growing in consciousness. To help that happen, it has generated many conscious things inside it, including us. It already communicates with us…in dreams, unexpected insights, visions, trance, and other altered states.

We can learn to communicate with it more consciously, we little beings who walk around on soil and pontificate about life's events and hardly notice that our universe is alive, nor even that our own planet is alive, even though what we call Mother Nature shows such intelligence in its clever organization of synchronistic events in environments tweaking for life, for animals and plants tuned to the day and night, to nutrients and locations, to wind and rainfall.

Amazing, too, that this universe realizes our current techno-culture does not yet perceive it lives, yet it is still evolving us and is opting for our future unless we humans opt to kill off our own chances and subside with a whimper.

If it is still impossible for you to imagine that something so large and diverse as the universe could actually be alive, at least toy with the notion that it is an integrated living system. Further, that it functions via polarized gravitational pulses operating at the ultra-tiny mobic scale, a scale even more fundamental and fine-grained than the quantum scale, which is already too small for science to see directly. This TOE proposes that at the mobic scale, a master code generated genetically identical cells that communicate and coordinate to common purpose within the universal body, which is alive.

In 1900, contemplating even the quantum scale required a revolution to expand the range of science, although it was just one more revolution after many before in science. When Max Planck studied the color changes in photons radiating from a black-body (it absorbs and emits all light frequencies), he was astounded to find that as the temperature went up or down, the photons gained or lost energy in fixed units or chunks. Planck named those quantifiable units the *quanta* at the *quantum* scale, from Latin *quantum* for "how much."

Due to the conservative, classical physics mindset of that time, finding energy in quantified units went contrary to everything that science then considered possible. Planck wondered how the strange emission of unitized energy at such a tiny scale could possibly be true: "My unavailing attempts to somehow reintegrate the action quantum into classical theory extended over several years and caused me much trouble." Planck himself suggested that he probably could not understand or embrace the revolutionary implications of what he'd found.

But it soon became clear to others that Planck had found the tiny scale where matter and energy emerge into existence.

2. Where did the original antimatter go?

Why do I think the universe is alive? Perhaps it started back in high school, when my friends and I asked, "Where did all the original lost antimatter go?"

Why did we wonder that? It's a roundabout tale. One of the guys, Mitchell Bronaugh, had a kindly neighbor he called Auntie Mattie. We loved her, too, since Auntie Mattie was generous with her cookies and her kindness. We even started an Auntie Mattie Club in her back yard to maximize the cookie factor.

Somehow Mitchell mentioned our Auntie Mattie Club to the high school physics teacher, Mr. West. Somehow he misheard Mitchell and thought it was an Antimatter Club we had started. Cheerful, bald, rabbit-toothed old Mr. West began to offer us articles and books on antimatter. We jokingly began to read them, or at least partially, and for a while, it actually turned into something of an antimatter club, with Auntie Mattie as its cookie-bearing sponsor.

In just such a club, without rebuke, the child mind may wonder about **Question 5: Where did the original "lost" antimatter go?** (It's on the *List of 20 Cosmological Questions* right after the *Introduction*.) Physicists think matter and antimatter must have originally been generated in equal quantities at the Big Bang. Why? They think it because if particles slam together at high speed in an accelerator, newborn particle-and-antiparticle pairs always appear. They mirror-reverse each other with flipped positions and opposite charges. Thus the particle and antiparticle pairs are mirror-reversed twins.

Normally they rush to meet each other and mutually annihilate by canceling each other out upon meeting. Both disappear in an explosion that releases an energy equivalent to their combined mass. Below you see two different particle-antiparticle pairs. Each reversing-mirror pair is busy at mutual annihilation, leaving only some energy waves fleeing the scene.

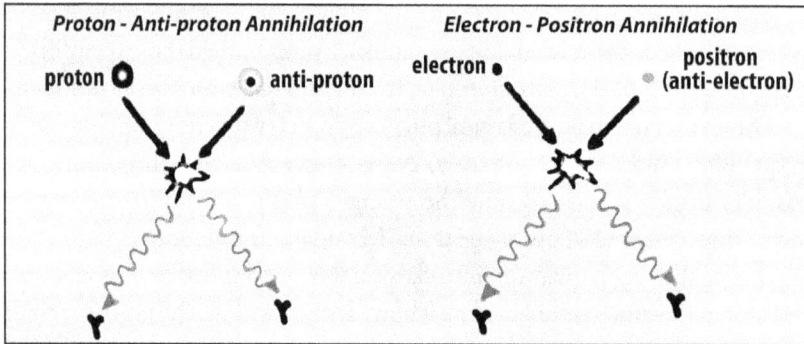

| Proton - Anti-proton Annihilation | Electron - Positron Annihilation |

proton ⦿   ● anti-proton   electron ●   positron (anti-electron)

Two particle-antiparticle pairs; each pair is self-annihilating

Physicist Paul Dirac, the discoverer of antimatter, explained mutual annihilation this way: a newly created antiparticle is like a hole dug in the ground. Its mirror-twin matter particle is like the pile of dirt sitting nearby. Their mutual annihilation is like the hole and dirt pile rushing back together, refilling the hole, canceling each other out to make a ground-level state again.

In a lab, this mutual annihilation always occurs when a particle-antiparticle pair are created unless scientists somehow manage to engineer a weird corralling of the particle and antiparticle into separate traps under extreme conditions, for instance, into two separate, cold, magnetic, or ultra-vacuum containers.

Scientists look around in the big 3D space bubble of our known universe and find that it has lost all of its original antimatter, although it still has lots of matter. What happened to all that missing original antimatter? Where did it go?

And why didn't everything just self-annihilate, canceling out in a Bigger Bang of energy that ended the universe almost as soon as it began? Why didn't it all subside back to a ground-level state, leaving a balanced zero of nothing?

Instead, here we are, you and I, living on this globe made of matter held together by energy. We look around and find that lots of other matter is sprinkled throughout the universe...yet any present-day antimatter that we find is explained by its creation in the ongoing events of stars, cosmic rays, and PET scans, where high-energy particle collisions are constantly taking place.

Although the laws of physics suggest that the original matter and antimatter should have immediately canceled each other out at creation, leveling it all right back to the zero ground state again, our universe somehow does exist. Matter somehow was left to hang around here and build things, while all the original antimatter disappeared. Why? What happened to "disappear" it?

Current physics views the disappearance of original antimatter as one of cosmology's greatest mysteries. It's been called the first great asymmetry...this universal lack of symmetry between matter and antimatter.

Some physicists speculate that when original particles and antiparticles rushed together to cancel each other out, perhaps only a tiny portion of matter was left over from that event—maybe only about 1 electron and 1 proton were produced from every 1 billion cooling photons—and that relatively miniscule amount of particles developed the universe we know. They say maybe any matter here now is just residual surplus left over from that cancellation.

Others say, no, maybe super-heavy, right-and left-handed antineutrinos existed right at the beginning, and maybe they decayed asymmetrically? Or maybe the asymmetry was due to a slight imbalance in the number of electrons and positrons? Or maybe it happened because meson particles that include quarks called *strange* and *bottom* annihilate asymmetrically?

Something is inadequate in each of those answers. None of it can account for the universal asymmetry between evident matter and antimatter A few resort to "Oh, maybe the laws of physics were just different back in the early universe." Or even "Perhaps the original antimatter loped off somewhere else to hide."

Physicist Maurice Goldhaber preferred that last idea about antimatter going off somewhere else to hide. Somewhere unknown. He said that the original antimatter is obviously not still here amidst us. Nor did it exist around here for very long. Otherwise, it would have rushed toward matter and gone BOOM! to annihilate everything in a Bigger Bang. Goodbye, Universe!

We know that BOOM didn't happen. We do exist. Goldhaber suggested the antimatter must have gone away or separated off somehow. He proposed there might be two separate universes of matter and antimatter, but located so far apart that they cannot possibly interact to cancel each other out. He called them the *cosmon* and *anticosmon*. Hmm, I like Goldhaber's terms of *cosmon* and *anticosmon*. It pings the idea of antimatter existing in separate realms.

3. The Double Bubble TOE came from a dream

This TOE agrees that original matter and antimatter do indeed exist, but they're not in two separate universes. This TOE says the matter and antimatter are not separated by mere distance. Were it only that, why then, given enough means and time and a fancy enough hyper-D ship, a crew might travel far enough to meet their doppelganger counterparts in the antimatter realm and mutually annihilate by canceling each other out, at least according to some science-fiction scenarios.

No, the two realms are separated in a manner far more radical, vital, clever, workable. I first saw it in a dream, back in 1985. I was teaching at the University of Texas at Austin when a startling dream took me into a union with the divine. Me, an agnostic! I saw the universe's originating dynamic. Me, a non-scientist!

Then I awoke, and rather like Planck, I could not begin to fathom the import of what I'd d experienced in that dream. At first, I only knew that it took me back to when our universe began, back to a place where mind and matter met. Then it went back before that, to an event of all the diverse universes beginning. It carried me into a place where science and spirit quit dueling to join hands in a deeper reality below their apparent contradictions. (The details of that dream are in *Double Bubble Universe*, Volume 1 of this series.)

In a nutshell, I began to realize the dream was showing me how our universe originated from what science would call nothing. The cosmic egg, or cosmegg for short, found an identity in nothing as number: 0. It began with a pulse of sheer being versus nonbeing as 0, 1, 0, 1. When it found the ultra-tiny mobic scale, the cosmegg began diversifying its pulses of being and nonbeing, and it eventually generated a single hourglass cell of 3D space and 3D time that projected on either side of the mobic scale. That success led it to generate many more such cells in the "nothing" around it, merging into two giant holographic bubbles conjoined at their membrane interface. Our science calls that event inflation.

Dimensional latticing projected on either side of the mobic scale, established by pulsing triplets that pair-bonded into 64 different 6-packs of polarity. Any pulses left over from making the upper bubble's spacetime lattice got repurposed into matter. But any pulses left over from making the lower bubble's timespace lattice got repurposed into antimatter. Physics sees broken symmetries from that process in the triplets and octaves of various particle-waves.

In other words, the cosmegg used numbers to work out a master code for a polarized pair of pairs—the space-time pair and the matter-energy pair. Or for short, spacetime and mattergy. Eventually, the master code that generated the Double Bubble universe acted as a template for a lesser variant, the genetic code that generated us. It used a quartet of molecules, a polarized pair of pairs: the Thymine-Adenine pair, and the Cytosine-Guanine pair. Their molecules spiral in your own DNA as triplets pair-bonded into 64 different 6-packs of polarity. Chemistry sees broken symmetries from that process as atoms organized in triplet states and in the periodic table's octaves of elements.

We are bits of organic life, generated and maintained by the genetic code. We live in this universe that is itself alive, generated and maintained by the master code. Just as the genetic code grew each species, updating it continually, so did the master code grow the universe, updating it continually.

There are many more parallels. Your body was developed by stem cells. So was the universal body. But unlike you, the cosmegg had to invent its first stem cell. I call it a **stem** cell because it originated all space, time, energy, and matter. That first stem cell of dimensionality was hourglass-shaped. Then it multiplied into many such cells stretching across both bubbles from the membrane interface at the mobic scale. Our universe lives. It maintains and updates all events in its body. What a miracle! Oh, consider its rocks as teeth, if you must, to get the idea that it is alive. Consider its water as blood, its suns as surging breaths of fire that come and go all over its body like pores breathing.

4. Two bubbles in one universe

To understand creation, go back to its origin. You can turn a balanced state of nothing into something by splitting its neutral equilibrium into two polarized states. Like this: 0 = -1 and +1. The polarized split creates not just one something out of nothing, but rather, *two* somethings out of nothing. This can be represented by a particle-antiparticle pair with mirror-reversed properties. They are twins with flip-flopped traits...

...as are a pair of human mirror-twins; their flip-flop features mirror each other's appearance bilaterally. In the next photo, many pairs of twins appear proudly at the annual Double Take Parade in 2010 in Twinsberg, Ohio.

Double Take Parade of twins 2010 - Twinsberg, Ohio

This sign shows bilateral reverse-mirroring, as if you're reading it in a mirror.

Bilateral or side-to-side mirroring

Recall, if a particle and its antiparticle are isolated into two separate containers, they can continue to exist in a lab. That's what happened to the cosmegg. Just as 0 can bifurcate into -1 and +1, so did the cosmegg bifurcate into mirror-reversed twins with polarized gravitation, spacetime, and mattergy.

The Double Bubble layout uses up-down reverse-mirroring...

Double Bubble

Up-down mirroring

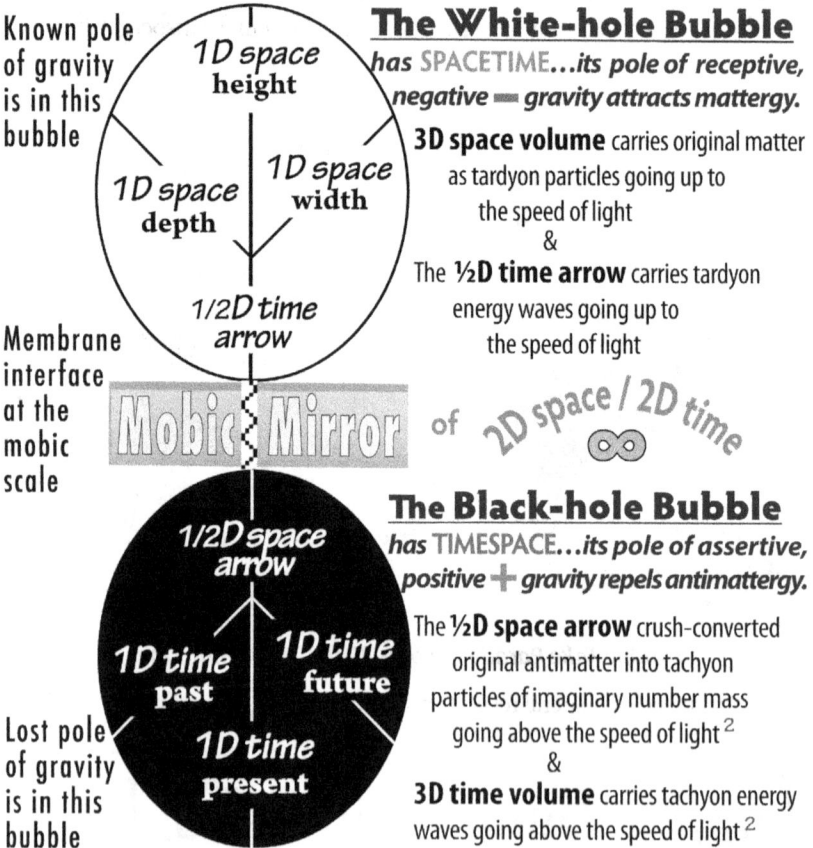

Known pole of gravity is in this bubble

1D space height

1D space depth

1D space width

1/2D time arrow

Membrane interface at the mobic scale

Mobic Mirror of 2D space / 2D time ∞

The White-hole Bubble

has SPACETIME...*its pole of receptive, negative ▬ gravity attracts mattergy.*

3D space volume carries original matter as tardyon particles going up to the speed of light
&
The **½D time arrow** carries tardyon energy waves going up to the speed of light

The Black-hole Bubble

has TIMESPACE...*its pole of assertive, positive ✚ gravity repels antimattergy.*

1/2D space arrow

1D time past

1D time future

1D time present

Lost pole of gravity is in this bubble

The **½D space arrow** crush-converted original antimatter into tachyon particles of imaginary number mass going above the speed of light[2]
&
3D time volume carries tachyon energy waves going above the speed of light[2]

The Double Bubble universe has up-down mirroring

This TOE says the mirror-twin bubbles have polarized, complementary properties. Each bubble holds one pole of gravitation, and that's why we here in the upper bubble know only gravity's attractive pole, not its repulsive pole. Our upper bubble is also polarized to hold 3D space and the ubiquitous arrow of ½D time, normal matter and energy, or mattergy, for short. However, that other bubble is polarized to hold 3D time and its ubiquitous arrow of ½D space, plus antimatter and energy, or antimattergy, for short.

Why is each bubble's arrow only ½D? Because each arrow only goes forward. Both omnipresent arrows are actually the upper and lower halves of a tensor network that constantly 8-loops across both bubbles, switching polarity from time to space or vice versa upon crossing the mobic-scale membrane interface.

In our bubble's 3D space, many tiny, diverse, electronic minds carried in mattergy bodies exist in constant *now*. But in that other bubble, a single, huge, unified tachyonic mind spreads throughout 3D time, existing in constant *here*.

Both bubbles communicate to cooperate along the mobic-scale interface, yet each bubble also maintains its own separate identity. Thus the conjoined twins interact, carrying on separate but related events, according to how each twin's specifications manifest the four primals of space, time, matter, and energy.

Thus this TOE says the original matter and antimatter got separated by something more profound than mere distance. The "disappeared" original antimatter is not gone. It is merely unnoticed, much as Edgar Allan Poe's purloined letter sat in a frame hanging right there on the wall, but unnoticed.

We take our own bubble's mental frame so much for granted that we've not yet opened our eyes to nature's bigger picture. We're oblivious to what exists just beyond the mobic scale: conjoined twins of complementary properties.

Yet if we open our vision to a larger frame of reference, we can realize that a conjoined mirror-twin bubble is tucked on the other side of the mobic-scale interface, existing in polarized complement to our own familiar bubble. That other bubble operates by laws of physics that are reciprocal to ours, yet seemingly bizarre because they are the opposite of what we experience up here.

5. Antimatter's vanishing act

Picture this scenario: at the great symmetry-breaking that physics calls the Big Bang, the burgeoning universe sorted itself out. It became two polarized bubbles, conjoined twins that interface at the mobic scale via a thin, mobic membrane. The two bubbles are made of four primals, a polarized pair of pairs: space and time; matter and energy. Due to the polarized constraints for each twin bubble, the four primals must appear in mirror-reversed formats.

1st Constraint: In this upper bubble, space can spread in three contiguous dimensions, but time is just a thin, ubiquitous, one-way, ½D arrow. (We call it *spacetime*.)Meanwhile, in the lower bubble, time spreads in three contiguous dimensions, but space is just a thin, ubiquitous, one-way, ½D arrow. (Call it *timespace*.) Each arrow is ½D, moving in only one direction: ahead. Together both arrows become a single polarized dimension whose myriad vectors move in a ubiquitous tensor network 8-looping across both bubbles, constantly switching polarity on crossing the membrane interface between bubbles.

2nd Constraint: In our upper bubble, matter travels in 3D space, while energy travels on the arrow of time. In the lower bubble, antimatter was long ago compress-converted by its narrow arrow of space into tachyon particle-waves...courtesy of $E = mc^2$, which holds universally.

Tachyon particles have imaginary mass (imaginary in the number sense, that is, as the square root of a negative number). Since according to this TOE, matter travels in space, but energy travels in time, upon conversion, that other bubble's tachyon energy escaped the thin ½D space to travel in capacious 3D time...whereupon that 3D energy powered up and evolved a huge, unified mind constellating throughout all 3D time. It thinks so fast because tachyon waves travel at a minimum of lightspeed[2], but usually at 6 to 8 times that fast.

3rd Constraint: One gravitational pole sorted into our upper bubble. The opposite pole sorted into the lower bubble. Our bubble's pole attracts its mattergy, but the other bubble's pole repulses its antimattergy, causing the lower bubble's 3D time flex-bag to expand gradually.

That other bubble is so near, yet very far removed from our normal senses and any means of confirmation by our current gross tools of physical measurement. Indeed, our five senses can only perceive and measure what exists above the much coarser grain of the quantum scale, where matter and energy emerge like dew coalescing in 3D space along ½D time.

That's why, when a lab creates a particle-antiparticle pair in this upper bubble, they immediately disappear again in an explosion. You see, antimatter is unstable up in this bubble. It cannot exist here very long because it is naturally polarized to be in that other bubble. So in a lab or a star, if an antiparticle gets birthed violently into this bubble's hostile parameters, it rushes to escape the wrong dimensionality with the wrong gravitational pole. It seeks a bubble that it cannot find, gawky in the unexpected 3D space and ½D time.

Frantic to rebalance itself any old how, the antiparticle finds a mirror-twin particle to adjust the lapse in nature's accounting. It enlists that reversed twin to redress the imbalance by mutual annihilation. When it intercepts the mirror-twin particle, they hug, then go *boom!* So long after creation, that is the only exit an antiparticle can find out of this upper bubble.

6. Guth said "repulsive-gravity material" inflated the cosmegg

In a 1999 talk, physicist Alan Guth proposed that a "peculiar repulsive-gravity material" caused Big Bang inflation. He suggested that right after the Big Bang, the existence of matter with a high energy density could have created a temporary "false vacuum" or "peculiar repulsive-gravity material" [with the trait of antimatter], causing a vast universal inflation.

To quote Guth again in a 2010 episode of *Through the Wormhole* TV series, "I came across this idea of inflation, the idea that gravity can, under some circumstances, act repulsively and produce a gigantic acceleration in the expansion of the universe, and that this could have happened in the very early universe. The key idea behind inflation is the possibility that at least a small patch of the early universe contained this peculiar kind of repulsive-gravity material. And all you need is a tiny patch of that, and the Big Bang starts to do this repulsive-gravity effect."

Various discrepancies, however, popped up in Guth's inflation model. Physicist Paul Steinhardt, for instance, was an early advocate of it, hoping it would resolve some Big Bang quandaries. But then he turned naysayer, telling Amanda Gefter in an interview in *New Scientist*, June 30, 2012: "We thought that inflation predicted a smooth, flat universe. Instead, it predicts every possibility an infinite number of times. We're back to square one."

7. How our Double Bubble universe avoided the bigger bang

Not quite. This TOE says Guth was nearly right on how inflation started. Antimatter was a factor, but not in the manner he suggested. Instead, the Big Bang blew two polarized bubbles. That first force's gravitational polarity sorted the original matter and antimatter into two different bubbles. We got matter.

Many clues in physics already do hint at our universe's Double Bubble structure. Scientists notice discrepancies for matter and antimatter regarding particle charges, parity, and time invariances. Odd discrepancies exist between the size and weight of almost all negative particles compared to their positive particles. Moreover, other vexing issues such as dark matter, dark energy, and a missing pole of gravitation whisper that our current experiential spacetime signature—3D space and 1D time, shaped like a single bird claw—sees only this upper bubble, when actually, two feet exist in precise counterbalance.

I admit that what I am suggesting here is not how current science explains puzzles like the missing antimatter, missing pole of gravitation, dark matter, dark energy, or the mystery of mind itself. Current science considers them all to be separate puzzles, and each a great mystery. Relating these mysteries and resolving them with one solution could show how they all point to a reversing-mirror bubble that sits on the other side of the mobic scale. Resolving those mysteries would make a convincing argument for this TOE, which proposes that our Double Bubble universe has a mirror-twin structure, and it was generated by a master code that was forerunner to our own genetic code.

Here's the crucial point regarding antimatter. Yes, physicists can create particles and antiparticles that annihilate each other in the lab. But the resultant annihilating explosions do not mean that antimatter must *inevitably* cancel out matter in all conditions, and thus it inevitably did so in the early universe.

It did not. Witness the possibility of keeping matter and antimatter in two separate containers. That's how the burgeoning cosmegg managed to divert its fate from total obliteration at a lethal intersection of all matter-antimatter pairs. It simply redirected a destructive *tete-a-tete* that could have occurred on a universal scale by establishing an infinitesimal-scale, mobic membrane interface between both bubbles, corralling mattery and antimattery into two separate traps under extreme conditions.

Mere polarized scaling in a membrane interface allows this Double Bubble universe to remain intact and functional as a whole. Indeed, super-functional, due to that other bubble converting its original antimatter into tachyonic energy traveling in 3D time. That super-swift energy could power up and develop the huge, unified, universal brain that developed us, which some call Mother Nature.

The other bubble, that conjoined mirror-twin, is what made our whole universe smart, able to survive and evolve. Its huge, unified brain communicates constantly at the interface with this other bubble where we live, and even to us personally in sudden insights, dreams, remote viewing, channeling, trance, and other altered states. It accounts for David Bohm's "Implicate and Explicate Order," and it explains why Bell's theorem works...why two entangled particles can influence each other instantly without being in each other's locale.

Certainly, yes, in a physics lab experiment, particle-antiparticle pairs will meet and explode, but that event happens much more often, and with much greater effect, in our sun. Moreover, a very massive star can go super-nova and condense into a neutron star, or sometimes even turn into a black hole.

So if it were not for that thin, mobic membrane interface of 2D space and 2D time between the reversing-mirror twin bubbles, this potential universe would have reverted to annihilation. But it didn't, and here we are now, taking a look at what I first saw in a dream back in 1985, a Double Bubble universe.

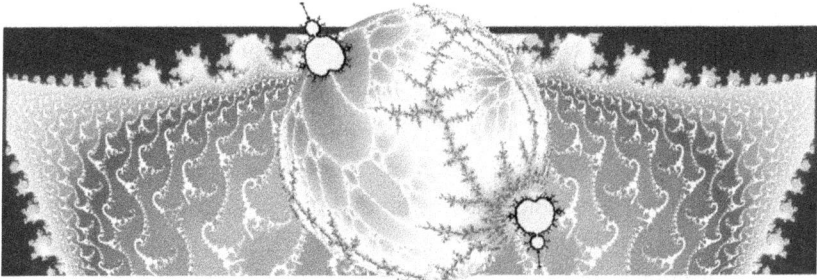

Chapter 2: Beyond the Linear Limits

1. The Copernican revolution

Objective analysis has long been the main feature of Western science. It offers much to inform science. But does objective analysis have its limits?

Objective analysis moved into high gear in the West as we adopted the Copernican idea that the earth is not the center of our universe. That upstart notion flouted the religious belief of European culture at the time.

De Revolutionibus by Copernicus

Nicolaus Copernicus was hesitant to publish *De Revolutionibus* because he was afraid of being charged with heresy. In 1530, he nevertheless began to circulate a summary of his ideas in Latin among a few scholars. When his complete book was finally printed in 1543, Copernicus considered himself fortunate to be on his deathbed and thus immune to execution.

De Revolutionibus fostered a mental revolution that eventually dominated Western science. Its implicit dictum: use logic's detached study of an object instead of getting mired in subjective relationship with it. Stay logically objective!

Why did we honor this new ideal so much? For the best of reasons...it worked so well. Within about 150 years, Newton's calculus could measure and predict the rate of change in known physical systems, which led scientists to assume that all physical phenomena could be understood by this method.

The Newtonian scientific method could determine if something is *valid* (it apparently approximates the truth) and *reliable* (the same truth is shown each time you look at it). Scientific certainty reached such heights that in 1814 Pierre-Simon Laplace claimed if he knew the position of every particle in the universe, he could predict the universe's entire future in all its details.

At the level of reality that our senses inhabit, Newtonian science has been very successful. It elevated living standards so much that objective logic became the touted criterion honored by modern culture. It championed the detail-oriented, goal-focused left brain over the holistic, network-oriented right brain.

Culture began reinventing social norms and standards—for example, by turning architecture into a cool model of linear logic, erecting straight-edge buildings that resist emotional relationship, unlike the old Greek or Chinese temples that applied a sly art of proportions that trigger your visceral response of "Ah, that's art!—not just engineering."

2. Light creates shadow

The touted scientific approach in Western society brought many benefits, but as patriarchal power used technology to commandeer wealth, governance, and now data, that yang power also downsized the constellation of yin values: equality, acceptance, patience, sympathy, humility, charity.

An increasingly scientific mindset considered the God concept to be objectively unprovable. Religion began to seem a mere projection of the naive desire for life beyond this frail cage of bones. Since meaning must finally remain subjective, many shed the notion that life has any larger meaning.

Following the new god of technology has made our lives easier, but not depending so much on others also weakened our blood, clan, and emotional ties, moving us gradually into disposable relationships and rotating stepchildren.

Lifelong loyalties have watered down to a flux of solitary TV dinners, temp jobs and residences, interchangeable acquaintances. Electronic access to goods, services, media, music, sex partners, news of the moment at *every* moment—it all hyped the message of an exciting yet meaningless, frantic yet empty life—reinforcing more isolation, insecurity, narcissism, and alienation.

As the vast forces of big corporations, big government, big money took control, high-level crimes against society were perpetrated by conflict-fanning politicians, Wall Street moguls, and the ongoing pollution of the environment. Books, films, and other media began to portray the common man as an anti-hero continually frustrated by finding no exit from bureaucracy's stale horrors. No relief from deception, corruption, and tyranny, living stripped of the frills of ethics, sentiment, and ultimate meaning in a bleak dystopia full of goods.

The fake-friends, fake-news, fake-fun society skewered its vast population on the chilly statistics of market share, internet theft, and continual violence. Various cults and anti-heroes committed mass killings and suicides, both literal and indirect, both fast and slow, often attended by a media blitz of coverage.

The yin virtues got sent to a cultural doghouse in Coventry, banished beyond the pale, off to Hell and gone. Put that weepy bitch in a chastity belt! Burn her as a witch! Keep that crazy aunt in the basement…where she gradually got rude, screwed, and tattooed. She turned into an abuse-toughened waif who played with fire and hornet nests, gesturing "Up yours!" at society.

Because, of course, Hell is not buried underground, but instead, it hides in the repressed yet constantly eruptive unconscious of an overly-selfish, consumption-oriented, goal-obsessed society. A macho, take-charge culture has no power to stop its own bloated shadow from proving again and again that it does not have everything under control. Not really. Not at all.

3. Scientific giants with starving souls

But during the 20th century, seeds of change were already sprouting. New "soft sciences" charted the myriad angles on a holistic reality becoming evident in the new global mix of languages, cultures, psyches, architectures.

Even the most respected of hard sciences—physics—was encountering a strange, new, and subjective fuzziness in its technical data. Einstein showed that space and time can only be measured relatively, that matter and energy are disguised versions of each other. Quantum physics found that light is both particle and wave, yet those two conditions seemed somehow not simultaneously and equally measurable. What! Why?

New geometries emerged to challenge Euclid's old theorems, new algebras updated the old laws of math called *al-jebr* in Arabic, meaning "reunion of broken parts." Foundational physics and calculus became limited, modified, or abandoned. British biologist Rupert Sheldrake had the temerity to speak of the *habits* of science, rather than the *laws* of science.

With new instruments, we began to chart realms far beyond the naked senses. Tools artificially extended our perception and allowed us to enter new macro and micro realities that were formerly invisible to us. We discovered with a shock just how limited is our sensory view of matter…energy…space…time.

Those newly developed technical sense extenders took us into places where the old laws of Newton's physics, and even his calculus, did not apply anymore. Data from beyond the sensory fringe opened a vast, new patterning where the old linear paradigm falls away. Here nonlinear dynamics whisper of unimagined meshings. A traditional scientific mindset cannot perceive or describe this new universe of holistic relationships whispering all around us. In us.

We are shocked to find that scientific objectivity does not, in fact, finally even seem to exist. It is a fond delusion in much of science. Our research is forever limited by the cultural and even species-specific mindset we carry. What we see and how we make judgments on it inevitably depends upon our human brains, on our varying chemistries, cultures, stances in the hologram of reality.

We've created gigantic, scientific extensions of our sense organs; we grew enormous, technologically enhanced eyes, ears, noses. Yet we're also beginning to realize that we have not kept pace in understanding the human psyche, that fancy Greek term for a blunt, Anglo-Saxon, four-letter word: *soul*.

We are still pygmies of the psyche. Science puts rockets and satellites in outer space, yet it often scoffs at the mysteries of inner space where mystics roam. Some still want to fit the psyche into an old mechanistic mold where linear logic should hold sway (but doesn't). Fortunately, Freud opened a royal road to the unconscious via dreams. Jung tracked archetypal patterns in dreams, and they can only be experienced personally, reported subjectively.

A paradigm shift is starting to take place in science now, and it requires a colossal adjustment. It is reorienting the Western mindset after 2,500 years of increasing emphasis on linear, objective logic. For a long while now, we have measured our reality in coffee spoons, inches, kilos; we have linked cut-and-dried units in cause-and-effect sequences to reach specific goals. True, it has worked very well at the daily material level described by Newtonian science.

But we suddenly find that we've gotten lost way down in the weave of scientific details, counting out steps, tracing narrow threads of logical linearity in a measured tread to various dead ends. The web of connectivity above and below Newtonian science now offers the wonder of observer-sensitive electrons and pairs of entangled photons that communicate instantly across distance.

Eddington was fascinated by the "Constants of Nature" that Newtonian science found inexplicable. By now, physics has a list of 26 dimensionless constants that allow the universe to exist. They're called *dimensionless* because they rely on ratios of relationship rather than on specific units of measure such as seconds, inches, or kilograms, and often their equations use $\approx$ for "about like."

4. The underlying paradigm is analinear

In the first three books of this series, I described how DNA's 64 molecular 6-packs correlate mathematically with the I Ching's 64 hexagram 6-packs. I explained how the trigrams in those hexagrams can shorthand the 64 RNA codons that automatically sort into their correct amino acid families and have philosophical I Ching messages that even reflect their amino acid tasks.

These two different systems of 6-packs correlate so well because they share the same underlying mathematical paradigm. Its basic, easy math has linear drive, but it also holds resonant networks of numbers that explore and amplify their analog relationships. It uses both binary units and analog resonances, and they combine in a nonlinear dynamic so special that I call it analinear.

I suggest that the underlying mathematical paradigm came from a master code that generated the Double Bubble universe itself. Its pulsing of 64 different polarized 6-packs of information developed the dimensional lattice projecting from the membrane interface between the white-hole and black-hole twin bubbles that exist as symbiotic, mirror-reversed partners. We live in the white-hole bubble.

This TOE claims that a vast, relational order moves deep within the apparent chaos of events. It says our universe is a giant fractal generator of the emergent flow that we call reality. Its patterned co-chaos cannot be fathomed by logical analysis alone. Instead, it moves without exact predictability, much as weather moves in iterating patterns of the seasons that can be predicted fairly well in their general forms but not in their specific details.

You and I move in our own iterating patterns within the whole, experiencing events that are often in similar patterns yet also uniquely different each day. The way that you find out what is going to happen today…exactly…is by living it. Sure, at times, your life may look and feel chaotic. But the constant in it all is you. Or me. Each of us acts as a focus for all of the shifting data.

Sometimes you may notice a pattern that is hidden within the apparent random details of your day. It emerges abruptly, triggering meaningful connections that are not only in your head, charting a discrete quantity of logical data, but also in your heart, vibrating with a relational quality of feeling. As the linear and holistic converge into a pattern, you may catch sight of the vast wonder behind a fleeting moment. These unique moments, when overlaid, multiplied, can even coax your life into the holistic flow of qualitative meaning.

5. A holistic realm that informs linear logic

If you do accept that there is a holistic realm of meaningful connectivity that can harmonize with linear logic and inform it, how do you start to explore it? Consciously, that is. Intentionally, purposefully. The main difficulty with the holistic realm is that it flickers at the edge of logic. It signals from beyond the linear boundary of consciousness. It keeps evading your direct gaze. Thus it must be approached with the sidelong glance of holistic pattern recognition, rather than with the knifing stare of a focused skewering and dissection by logic. Much as your peripheral vision can perceive a faint star with a sidelong glance when your direct gaze cannot, so does the liminal, sidelong glance of awareness beyond the ego limits discern enduring truths that the ego's focused stare cannot face head-on, and thus will not agree actually exist.

Jill Bolte Taylor in *A Stroke of Insight* tells of being dumped into the holistic realm by a massive stroke on the left side of her brain. She beautifully describes the euphoric, intuitive nirvana that arose via her right brain after the stroke. She felt a sense of complete well-being and peace, as told in her TED talk.

But how can you go there without a lucky stroke, a revelation, or a life-changing realization? My impetus was a life-changing dream. I saw the Double Bubble universe in a dream that changed my life as I began to realize the shape and weight and thrust of its import. That dream took me, an agnostic, into a grand organizing design so huge that I experienced it as God. Me, an agnostic!

I had to understand what the dream showed me, so I quit teaching at the University of Texas at Austin and went to study for 5 years in the Carl Jung Institute in Zurich. Then I went to China and taught at Jinan University in Guangzhou for a year so I could study with an I Ching scholar.

How can you prepare to enter your own unknown territory? How do you go about mapping its secret patterns? How difficult is it to discern and grasp the profound experiences that await hidden beyond your ego boundary?

Many cultures have historically left that to mystics, gurus, and prophets. They cross an elusive threshold of consciousness in trance, insight, meditation, dreams, fantasy, hallucinogens, whirling dervish momentum. They flow bodiless into a dark mystery and sometimes drag back a revelation to the light of day. That experience opens a door between the known and unknown to reveal some design beyond normal vision. Still, it is always tinged by the filter of that person's own personal stance in the hologram of reality.

You may pursue an insight glimmering at the edge of a dream, an uncanny coincidence, a Freudian slip, a Jungian pun, even a tarot or palm reading, if that is what your unconscious requires for permission to slip beyond the ego's boundaries into a place where information is real, yet you cannot quantify it.

Perhaps you can do it just by learning to grab your dreams before they slip away, instead of letting them evaporate like the foolish capering images of a pointless film. It means working to discern in them the archetypal framework of your own life. It means discovering how to use your ego as a handy tool and when to put it down and stand relaxed, empty-handed before the larger wonder of life to receive its gifts. Achieving such ease takes some effort.

We are now discovering—rediscovering—that holistic meaning is coded into the wild stories that arise from the unconscious in our sleep. When a dream's forgotten language is recovered, it becomes no longer "garbage in, garbage out." Instead, it offers you wisdom beyond the ordinary ego boundary. How do you learn to go there at will and with purpose? The ego absolutely resists that trip, knowing it will not remain in control within the holistic unconscious.

6. Mapping the hidden terrain

Beyond the ego boundary, you will have to face truths you probably would rather not see, and fear may project a frightful face upon those truths. To a rigid, brittle, impermeable ego, the resonant web of analog insight can be especially frightening because a logic-driven ego expects to find only demonic deception beyond its limits…and such an attitude can become self-fulfilling.

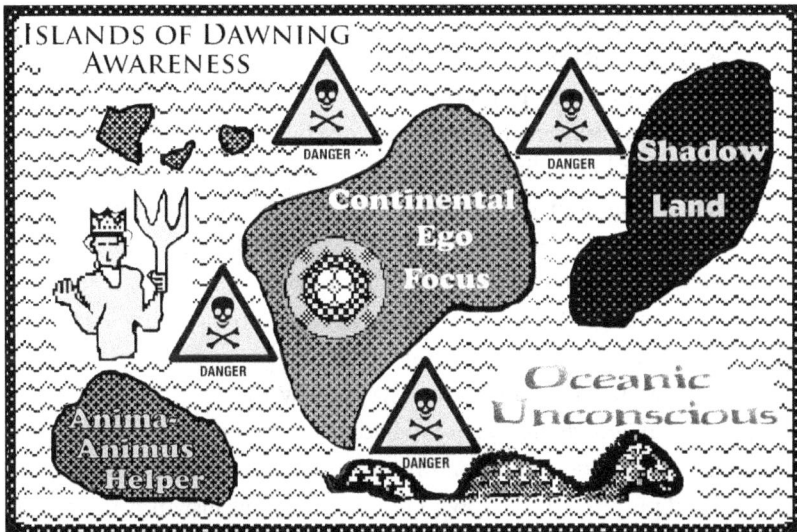

Ego speaks: "Beyond these gates there be dragons"

On the psyche's metaphorical map, the ego posts danger signs at the edge of its protected boundary: "Beyond these gates there be dragons." Nonetheless, ego also yearns to sail past its known limits into the mystery beyond, hoping to find adventure with benefit. The ego's exploration of islands of dawning awareness is difficult enough, but turning ego into a tool to see the Shadow Land and truly seek greater good for both oneself and others? That is harder still.

But don't worry about having complexes. In Jungian thought, a *complex* is just an organizing nucleus of behaviors and emotions. In itself, it does not imply good or bad. Complexes are not logical; they are relational, analog, networking. They're merely patterns that exist, and you can winnow, plump, and prune your complexes to render them more useful rather than detrimental in your life.

Your life's patterns create a unique pointillist painting made by the raw data speckling your days and nights with events. Sometimes you see a linear logic in its patterns, but often, you do not. Nevertheless, a deeper, underlying dynamic exists in your life, invisible to the ego. This TOE says the flow of daily events has its own hidden fractal dynamic, and you can often spot patterns hidden in the shifting flow. And change them.

This hidden flow of the Tao, the watercourse way, isn't visible to our physical senses…which doesn't mean it doesn't exist. You don't call gravity unreal, even though you only see its effects, not gravity itself. Did a caveman ever stop to wonder what kept his feet on the ground? Or why his spear dropped short of the animal and his intention? By now, we Earthlubbers know many ways to explain gravitation, but we still wallow daily in its dynamics.

Just as physics has learned to describe much about invisible gravity's visible effects, psychology, philosophy, spirituality, and mysticism are limning in more and more of the unseen geography that the psyche travels…which still extends beyond the linear, logical limits we have managed to describe in our sciences. That mystic realm is sourced best through *ahas!* of recognition—but it is quite a task for logic. No wonder Harvard-trained neurophysiologist Jill Bolte Taylor needed a stroke to usher her into the brokedown palace of holistic splendor.

The path through this series sometimes passes into the nearly inarticulate right brain, where holistic timing and spacing give us a glimpse of huge, analinear dynamics that can be shorthanded by the ancient I Ching's 64 hexagrams in an underlying co-chaos paradigm that is encoded in nature itself.

Enlarging your angle of vision on the universal hologram can reveal the meaning and purpose of your own life. Accessing this insight is beyond the grasp of your five senses, nor is it just following a knee-jerk, emotional reaction. Wisdom is different from knowledge and difficult to haul into the harsh spotlight of ego awareness…and harder still to articulate. No wonder Lao Tzu began the *Tao Te Ching* thus: "The Tao that can be spoken is not the Tao."

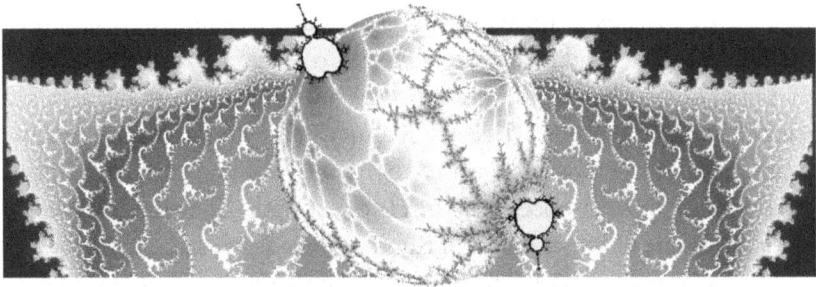

Chapter 3: A Stroll in the Dimensions Museum

In this chapter, we probe 🧩 **Question 6: What are dimensions?** To find out, we'll stroll through the Dimensions Museum, wearing a headphone guide that explains the history of dimensionality. Then we'll go outside on the terrace to view nature widening in a new, larger perspective on space and time.

At the museum entrance, we don headphones and hear a chirpy voice say, "What are *dimensions*? That word is often used loosely!" We pass a poster of a jazz concert as the voice gushes, "John Coltrane added a new *dimension* to a sax solo!" The next poster shows a cook measuring salt, and the voice exclaims: "Salt takes food to a higher *dimension!*" A third image shows people sitting in yoga's lotus pose, and the voice says, "Meditation explores unseen *dimensions.*"

We walk onward as the voice says, "You've just heard some very loose applications of the word *dimension*. The Dimensions Museum will explore only the dimensions that are physical extensions of space and time."

We enter a dark passageway where a new, deeper voice intones, "Space. Time. We take them for granted. We cannot directly hear, taste, or smell raw space. We cannot grab sheer time. But we experience them constantly in our everyday lives. Nevertheless, we can only see what is *filling* the space, only notice what is *occupying* the time. But sheer space and raw time themselves? They remain invisible to our senses.

"We can, however, examine them with our minds. Space is easier to examine than the mystery of time, so let's start there." A sign points ahead to the *Space Place*.

1. The Space Place

We enter the *Space Place* and walk around, looking at exhibits that show different ways to define space, measure space, occupy space. We see teaspoons. We read plaques about Land Wars. We watch moon rocket videos. There are lots of images and objects that show how we take up space, use space, conquer space.

We wander into a video alcove where a cartoon child is exploring space by moving one hand side to side...up and down...forward and back. The voice says, "Space extends in three dimensions—width, height, and depth. This child can wiggle a hand in all three dimensions at once. So can you. Try it."

I hold out my hand and wiggle my fingers, watching them move in 3D space. The voice explains, "This act does not pull your hand apart because it can move in *contiguous* 3D space. Your hand's volume occupies all three extensions of space at once. Length. Height. Depth. They mesh smoothly like a threesome of horses pulling a Russian troika sleigh across the snow."

We see a video clip of three horses pulling a sleigh quickly over snow. "This accord among all three space dimensions lets your hand wave around in 3D space. It lets you work in a 16-story building. It lets you do kung fu."

Then the voice asks, "To paint a room, how much paint do you need? You can enter it and calculate how much paint is needed to cover the walls and ceiling. You make a 90° turn"—a video shows a stick figure doing this—"and measure the room's length. You make another 90° turn and measure the room's width. You make yet another 90° turn and measure the room's height. Now you've run out of new 90° angles in this room's space—just length, width, and height."

The voice continues. "Euclid's breakthrough geometry was a great adventure in thinking. It expanded a dimensionless point to a 1D line, then to a 2D plane, then to a 3D solid." As I watch, a cartoon hand sketches a line that turns into a triangle, then into a tetrahedron. The hand stops. "Oops!" the voice says. "Here we bump up against the functional limit of spatial dimensions. We experience only three of them. We have nowhere else to turn.

"Euclid told us how to measure an expanse of 3D space, but not what the space is made of, nor how it is generated, nor why space exists. Space is your invisible friend that you cannot explain to anybody, yet you know is there. You just know it exists! Really! Why? Because you live in it, walk around in it.

"If space holds everything in this universe, why do we know so little about its nature? We should study it. Here's a proposal. Scientists routinely produce antimatter in labs to study it. Right? So let's make some new space in a lab and study it. Oops! Turns out, nobody knows how to generate new space in a lab...but we have learned a great deal about how to measure it, fill it, use it."

We leave the alcove and walk past more exhibits of space-filling items. They demonstrate how space gives us an invisible but dependable real estate to build our lives in, how it lets us move around in its commodious, contiguous, 3D volume that holds buildings, trees, streets, us. So yes, this *Space Place* is great!

2. The Time Tunnel

We go into the *Time Tunnel*. Wow, this narrow room makes you slow down and think. Here's a cartoon of Einstein riding on an arrow of time. A voice-over says: "Unlike space, time will not let your fingers move in two of its dimensions at once. Much less in three. You always experience time as right *now!* And right *now! Now!* Time shoots you along on its moving moment of *now!*

"Invisible walls sequence you again and again into the ever-present moment of *now*. You cannot leap back into an old moment of *now* except in memory. You cannot jump ahead except in anticipation. Time pinions you to the eternal *now* rushing headlong toward the future.

"You move forward, never backward, yet you never quite reach a future beyond *now*. And although you cannot escape this eternal moment of *now,* the *now* somehow keeps changing. It holds new events. Why? Why does *now* shoot ahead on the arrow of time *to* leave history trailing behind? Each life span—whether of a galaxy, a person, or a fruit fly—rides on time's relentless arrow. It shoots forward and leaves the past behind, dead and gone. You can go home again in expansive space, but not on the skinny arrow of time.

"Why can't you just leap into next Monday if you prefer that schedule? After all, you can enter the space of any room in your home. But you cannot experience any other part of time but *now*. No matter where you live, whether it be in various time zones, or even in a space ship orbiting above all the Earth's time zones"—we see illustrations—"it is always *now*. Somehow dimensionality lets your body cavort in 3D space, but confines it to the prison of ever-emergent *now.*

"In fact, all of the individual *nows* can be in radically different places, even moving at different speeds"—we see illustrations—"but in each radically different place, it is always perceived as *now*. No matter how gravitation affects you, or acceleration, which is its functional equivalent, experiential time is *now* for everyone…everywhere.

"Ponder the diameter of time, so thin that it has no circumference. You can measure only its length. Here's an idea! Why not build some time in the lab and find out why constant *now* keeps us suspended between the two points of *past* and *future*? Oops! Turns out, nobody knows how to build time. They just pretend to put it into a cosmetics jar, save it in frozen dinners, time-share it."

You and I raise our eyebrows, grinning at each other. Then we walk on through the various displays that show how people at first marked time by just surviving challenges. It apparently was hard to think past the precarious *now*…

…until people began to think of time as cycles. Sun cycles. Moon cycles. Here comes the harvest…again. Here is birth again. Death again. Let's plan! The human journey through time has developed mythologies and rituals to honor Earth's cycles of time, helped us schedule to plant seed, make channels to irrigate crops, harvest them, and build granaries against an uncertain future.

Exhibits show how people began to measure nature's cycles by setting time into circles. On a sundial, if a shadow moved from here to there, an hour had passed. That convention allowed cycles in time to be measured by an object tracking circles in space. Clocks appeared, and circles seemed to describe time…for a while. But the windup clock turned into the digital clock, atomic clock, pulsar clock. As we tried to master nature's cycles with timing agendas to control agriculture and stimulate trade, we began to view time as shooting for goals—linear—so by now, physics likens time to an arrow, not a circle.

For many folks, time moves ahead like an arrow, but it also repeats in cycles. That spiraling facilitates years of easy storage in the memory…like coiling a rope onboard…or like DNA's double helix coiling in our bodies to store our heritage.

We come to a *Sci-Fi* alcove that shows us amusing movie clips about time travel. The plots send people not to *elsewhere*, but to *elsewhen*. They travel backward or forward in time to visit other historical periods. They meet their ancestors, descendants…or even themselves, to nearly cancel out like a particle-antiparticle pair…even though all of that stuff never, ever happens in real life.

We see a clip from a sci-fi movie called *12 Monkeys*, watching it at least long enough to realize that fancy editing can cross-cut the past, present, and future to keep us suspended in a provocative series of cascading revelations. The guide claims these sci-fi stories help us reflect on the mystery of time. I suppose so.

We stop at the *Cafe Bay* to sip and chat. Here we both agree it's no wonder that space and time seem so different. One has 3D scope; the other keeps disappearing. Space stands available; time outruns each event. Space and time just don't look alike! Yet science claims they are dimensionality's two flip-sides. Why?

3. The Relativity Reach

We look for an answer in the *Relativity Reach*. A wall chart shows a timeline of when science started calling time a dimension. In 1827, August Mobius said a fourth dimension lets a 3D shape rotate onto its reversing-mirror image. In 1854, Bernhard Riemann established a geometrical theory of higher dimensions, especially four dimensions. But only in 1915 did people really began calling time a fourth dimension, when Einstein expanded his theory of special relativity into a more general theory of relativity. It was a geometric theory of gravitation that said gravity is just the curvature of time and space caused by matter and energy.

Einstein's mentor, Hermann Minkowski, already had a 4D spacetime theory. It combined 3D Euclidean space and 1D time in a manifold of 4D spacetime. To ratify his own idea of locating things in a 4D spacetime manifold, Minkowski persuaded the reluctant Einstein to call time a full dimension.

British astronomer Arthur Eddington wanted to test Einstein's idea that the actual cause of gravity was space and time warping around mass.

Eddington's 1919 photo for Einstein's general relativity

To do it, Eddington made an expedition to Africa to record and measure the displacement of starlight during a full solar eclipse. He brought new fame to general relativity in 1919 by managing to develop two usable photographic plates. His two photographic plates (good) and his calculations (bogus, some said) verified that our sun's gravitational field altered the path of a star's light.

Then in 1927, Eddington formally proposed that time moves like an arrow. (The idea had already long been in literature.) Physics soon found a whole quiver-full of time arrows at work. We walk past their showcases, reading the labels and watching the displays of various time arrows: causal, cosmological, thermodynamic, radiative, weak nuclear, and quantum.

Aha! I recognize this next exhibit! It's scrawled on an old-fashioned blackboard: a trident with three prongs of space and a handle of time.

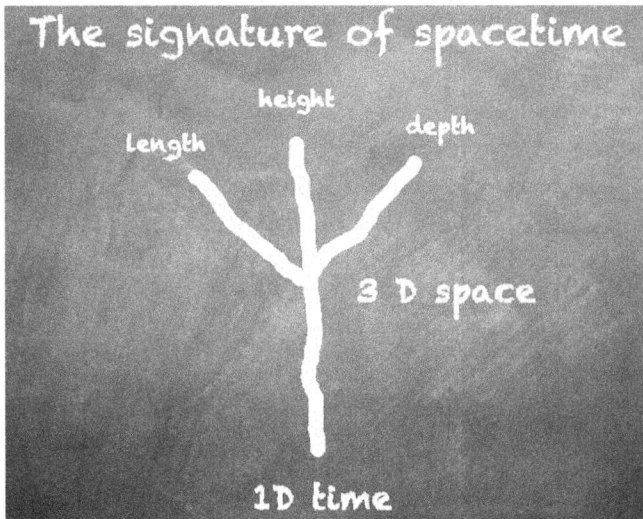

The standard spacetime signature

I exclaim, "This is the same diagram that my high school physics teacher drew on the blackboard back in 1957! Mr. West called this trident the universe's spacetime signature. He said the prongs represent 3D space and 1D time."

You laugh and say, "But that's not a trident. It's a bird foot."

"Hmm...I see that," I respond. "But in that case, this poor bird of spacetime has only one foot to stand on. And with just one foot, it limps, crippled and lopsided. Plus, even that single footprint isn't symmetrical. See how its talons of space and time count out differently?...3 for space, but just 1 for time. Which just makes it so obvious that 3D space is not symmetrical with 1D time."

You retort, almost smugly, "My college physics prof mentioned something about the asymmetry of space and time. He said of course physicists notice the imbalance of dimensions, but they don't really consider it significant."

"Not significant? But a central feature of physics is its many symmetries! Physicists have made some great discoveries just by postulating a hypothetical symmetry, then looking for it and finding out that it really does exist in nature."

You reluctantly agree, "Okay, symmetry is quite important in physics. I remember Nobel Laureate Steven Weinberg saying in *Dreams of a Final Theory,* "It is generally supposed a final theory will rest on principles of symmetry."

I mull this over. "So why hasn't physics found a perspective that brings this lopsided bird foot signature of spacetime into better balance?"

You point me over to a wall chart that diagrams how to make the universe's experiential dimensions become symmetrical...sort of like cosmetic surgery, I guess. But they perform an operation on time alone, not on space.

The idea for making time symmetrical goes like this: if the arrow of time goes forward, hypothetically, it could also run backward along the same track, like a train running in reverse. Or like a video being reversed. Reversing time would put history into reverse gear and run events backward. Nothing in today's laws of physics actually makes that impossible. Theoretically.

I say, "They make it sound like the time train would be an easy thing to turn around on its track if they could just break into the cosmic roundhouse. Never mind that we can't really run our lives backward and forward like in a movie. It never happens except in sci-fi. We just ride on the time train to wherever it drops us off."

You say slowly, "But even if we could switch time back and forth, make it go in both directions"—you wave your hand back and forth—"like waving my hand across 1D length in space?...well, wouldn't time *only then* become truly 1D? So isn't our one-way time actually just ½D? Just half a dimension?"

I nod, saying, "I get it! Look, even if scientists did prove that time is 1D by managing to reverse it, 1D time still wouldn't be dimensionally symmetrical with 3D space! And you know what is left completely unaddressed here? Finding a contiguous 3D time that parallels our contiguous 3D space!"

We both agree that the idea of a one-footed spacetime bird is inadequate. I point out a display where physicist Eugene Wigner says as much, warning that a simplistic time-reversal operation does not correspond to finding a true symmetry for space and time overall. It does not render time truly symmetrical with space.

Then you shake your head. "And you know what? All these theories ignore what people experience constantly! The blatant asymmetry of space and time in our everyday lives! I can point my finger in any direction of space, but only ahead in time. Why don't they examine the why of that? Instead, they just hypothesize these 10, 11, or 26 tiny, rolled-up, hidden space dimensions! And unseen superstrings vibrating in spacetime foam at the quantum scale!"

I say, "Okay, I'll bite. What is spacetime foam? I missed that exhibit."

4. The Amalgamated Super Crew

You nod me over to an alcove as you explain, "Physicists can't see into the tiny quantum scale. But they assume spacetime foam exists down there, and along its bubbles run strings of vibrations. Those strings are the basis of string theory."

"What! String theory is about spacetime foam?" But the headphone voice is already speaking now, explaining the idea of spacetime foam, saying it stems from Paul Dirac's idea of an infinite quantum sea. Einstein wondered what was waving down there on the quantum-scale sea that could carry a particle?

Sure, it's easy enough to look at an ocean and see what medium is waving and carrying a boat: water. But what invisible medium is waving down at the quantum scale and carrying particles? Einstein thought that spacetime itself must somehow be making waves that vibrate to create particles, sort of like how violin strings vibrate to make musical notes. Moreover, Einstein thought those waves were likely vibrating due to gravitational shifts at the quantum scale.

A cartoon begins playing on the alcove wall. Its title is *The Amalgamated Super Crew Explores Spacetime Foam*. As it starts, a camera pans down onto tiny bubbles. It pulls back to reveal a cartoon physicist in a tweed suit, sitting in an old-fashioned clawfoot bathtub, sort of like Archimedes in a bubble bath.

He speaks to the animated physicists standing around the tub: "Super Crew, as your leader, I, John Wheeler, declare that space and time bend and warp into a cellular froth here at the quantum level. This is spacetime foam, and each bubble is about 10^{-35} meter across. Smoother than shaving cream."

Jack Sarfatti leans over and inspects the foam. "But it looks so dirty! And it's full of ultra-tiny black holes! Wait! Aren't half of them actually white holes?"

The Princeton String Quartet begins to play a superstring ditty called

HO and HE. Sarfatti listens and grins, "Sounds sexy. Downright heterotic."

Michael Green bends over the tub to inspect the foam and says, "It's a perfect 496!" John Schwarz chortles and exclaims, "Cancels out the anomalies!"

Ed Witten chides from the other side of the tub, "Don't forget E_8!"

A huge bird waddles up. "Ed, are you Cockney? 'E ate what?"

Witten draws with his finger in the spacetime foam. "No, I mean $E_8 \times E_8$."

The big bird says, "Then 'e ate a lot!"

Wheeler says, "I suggest this spacetime foam is created by the energetic rotation of virtual particles. Their points are like garments churning in a washing machine full of fluid spacetime. Their gyration is what creates the foam."

Leonard Susskind protests, "No, particles do not resemble points. They're more like tiny rubber bands vibrating in many modes, whipping up the foam."

Balancing carefully on both feet, the big bird sighs. "Uh, folks, I'm the Double Bubble bird, and you are not at the tiny quantum scale where particle-waves emerge. No, you're way down at the ultra-tiny mobic scale where space and time emerge. This is where polarized pulsing generates dimensionality."

"What!" "No way!" "Really?" Consternation, questions, upset faces.

"See, what you physicists call the whole universe is just one bubble of my two-bubble body. I'm made of conjoined, mirror-twin bubbles. Right now, you're at the mobic-scale membrane interface that conjoins them. Here is where polarized pulsing made space and time. You live in my upper bubble. Up there, leftover pulsing became ionized plasma that made your space opaque at first. Then the plasma cooled and coalesced, and your space became transparent. Well, density fluctuations put some hot and cold spots in your microwave background."

Janna Levin exclaims to the Double Bubble Bird, "What? You're saying particle-waves are just dimensionality's leftover pulses that got dumped into our 3D space? So the original plasmic mass was just some recycled trash? You're saying *that's* how the universe got its spots? That's big news!"

The Double Bubble Bird answers, "No, the big news is this. You guys omit one of my footprints! You're blind to half of my stance on space and time."

The big bird keeps on talking, but I cannot hear it anymore due to all the arguments going on. I whisper to you, "Did we miss the big bird part of this dimensional museum?" You shrug, then nod, so we agree to go look for it.

But the others, all those scientists, are now trooping on into the *Particle-Wave Nave* next door. Okay, if they've decided to join us on this tour, I suppose we should reciprocate and stay with them. We agree to trail them in.

5. The Particle-Wave Nave

Double Bubble Bird is saying in the Particle-Wave Nave, "Take a look around, everyone. Think it over. Take your time. You need a lot more time."

Strolling in the long, narrow nave, I begin to groan. All these quantum showcases are filled with so many little details! Let me just summarize. Quantum field theory combines several lesser quantum theories to say that all particles are so-called *excitations* arising in underlying energy fields…sort of like musical notes carried on waves.

Over decades, meticulously confirmed experiments managed to unify the last three great forces to emerge—electromagnetism, the strong nuclear force, and the weak nuclear force—into a Grand Unified Theory or GUT.

But oops! Gravitation, the first force to emerge, still could not be integrated into the GUT in a meaningful way, so quantum field theory stagnated…

…until its unfulfilled conjectures morphed into string theory, which tried to explain gravitation in a way that would unite all four forces. Seeking to make string theory work, theorists hypothesized what they call "dimensions"— actually, they're just variables—10 dimensions in the 5 most viable string theories, and up to 26 dimensions for the open or closed bosonic string theories.

Now, get this…all but one of those many "dimensions" (variables) relate to space, not time. And it's mostly just space rolled-up so small that you can't buy real estate in it, or play baseball in it, or even store an atom in it. Various theories use Lie groups that posit 64 standard model fermions. Some have 8 × 8 variables; some have 2 × 32 variables. In *An Exceptionally Simple Theory of Everything*, Garrett Lisi described a new way to use the $E_8 \times E_8$ format in a vast geometry of dimensional spins, but again, it's all about space, not time.

To physicists, it became clear that they'd cooked up way too many theories for the need, and they all began to look like special cases of a more fundamental theory. Ed Witten managed to merge the five main theories into M-theory, joking meanwhile about what M meant…Membrane, Matrix, Murky, Magic, Mother, Mystery.

Summing it up, Nobelist Steven Weinberg declared, "Evidence has appeared in the past few years that these five string theories (and also a quantum field theory in eleven dimensions) are all versions of a single fundamental theory (sometimes called M-theory) that apply under different approximations."

The M-theoreticians dusted off or newly formulated some baroque algebras, geometries, and knot topologies, and their elaborate calculations came up with some very pretty math that could vibrate to beat the band and calculate proliferating universes forever…

…but with one big problem in all of that theorizing, and it was insurmountable. Their calculations gave an infinite number of possible answers, but nobody could tell which answer was right. Only one solution would deliver the goods of this specific universe. Maybe. But nobody could tell which answer would do it, or even if one actually did. There was no way to pinpoint a sole correct solution that made our particular universe pop out like instant flowers.

As we're examining all the dusty showcases, I notice that the Double Bubble bird is grinning gleefully at us. Meanwhile, the voice in my headphone is saying that no instrument yet invented can get down into the quantum scale and see exactly what happens when matter and energy emerge.

You say to me, "Hey, I'm ready to exit this dusty room!"

But behind us, the Double Bubble Bird is announcing loudly, "Ahem. Everyone, please listen." We turn around. Dibby says to us all, "Do you notice how all this juggling of particle-waves, spins, charges, and myriad so-called 'dimensions' in various theories ignores one critical clue?"

Several people respond: "Clue? Where?" "What did I miss?" "What clue?"

"You're ignoring the lop-sided stance of your bubble's dimensionality. A single bird-claw of 3D space and thin arrow of time! That one-footed stance is wonky, no matter what. But look at me. I've got two feet!" Stomp! Stomp! I grin.

"Sure, the proliferation of string theories merging into M-theory opened up new windows on particle-waves and dimensionality. But remember, what they call 'dimensions' are just hypothetical variables set up in Kaluza-Klein style to account for all the particle-waves and forces. They're not real dimensions.

"And none of that explains the fundamental generation of space and time, nor does it bring your science's lop-sided stance on gravitation into balance. I can show you more about this out on the terrace, at my nest in the p-tree."

I exclaim, "You want us to leave the Dimensions Museum? For a petri dish?"

"No, for a p-tree. A polarized bifurcation tree. Actually, since the p-tree also has roots, it is a double p-tree, or dp-tree. We must go out onto the terrace for you to see it. And to get there, we pass first through the *Gravity Chamber.*"

6. The Gravity Chamber

Everybody moves into the *Gravity Chamber*, where the Double Bubble Bird says to us, "If you look around in here, you'll realize why your physics needs a bigger, more fundamental theory of gravitation. Nothing you've done yet goes deep enough, wide enough to cover all the bases."

You and I walk around, looking into a multitude of showcases and entering various video nooks. In one, a cartoon Einstein rides a rocket at light speed to discover general relativity. In another, a cartoon Planck stares at particles emerging in quantified units, shocked to discover what he names the quanta. A cartoon hand starts painting on the floor...*Planck level*...but Planck shakes his head, horrified. "No, no, call it the quantum level! Name it for a principle, not a person."

In the next video alcove, many cartoon physicists are arguing heatedly in a lecture hall about gravitation. I say, "No wonder this is called the Gravity Chamber. It's a grave situation in here. This whole group can't agree on what gravity is or why it exists. They don't know how, when, or where it starts. They need a reporter to come in here and do a story that spells out gravity's Who, What, When, Where, and How."

You say, "Okay, here's the basics. Science has three main ways of describing gravity: big-scale, small-scale, and emergence. Einstein's big scale theory of general relativity is the most familiar approach these days. That theory says gravity is caused by spacetime curving around objects. The bigger an object—Earth, for instance—the more spacetime warps around it.

"As for the small scale"…you beckon me forward…"come see this display on Loop Quantum Gravity. It's full of tetrahedrons and curved space loops testing to see if geometric quantities like area and volume have quantum properties."

I come over and look into the case. "Tetrahedrons. At least they're familiar. So all this is about quantizing gravity? And to do it, they're quantizing geometry?"

"Yes. You have to follow certain specifications to make a tetrahedron. So in effect, you're holding the geometry of space to a building code: **preserve those tetrahedral angles, no matter how squashed their shapes get.**"

I say, "Look at this one! It's signed like it's artwork: *Rovelli & Vidotto*."

You answer, "But read the placard. It says this display case could only happen because Juan Maldacena set up some ways to parallel quantum gravity theories with quantum field theories. That allowed the two seemingly disparate views to correspond in their results. So their two different angles of approach merge into a so-called 'holographic duality' that renders the theories actually equivalent."

I say, "Huh? So the two different views made a hologram?"

"Not a real hologram. 'Holographic duality' is just a metaphor for how these theories' two different approaches reflect each other in predictions that correspond remarkably well. Here, this kinda sums it up." You point to a metal plaque from Cambridge's Center for Theoretical Cosmology. It reads:

"The adS/CFT correspondence states that the physics of gravity in 5D anti-de Sitter space is equivalent to a certain supersymmetric Yang-Mills theory which is defined on the boundary of adS. The adS/CFT correspondence states that the physics of gravity in 5D anti-de Sitter space is equivalent to a certain supersymmetric Yang-Mills theory which is defined on the boundary of adS."

"Gobbledegook," I sigh. "Maybe all these theories are special cases of a more fundamental theory, and they're all just coming at it from different angles. It sounds like they need to dig deeper to find a more comprehensive theory."

You say, "I think that's what Erik Verlinde is trying to do right now. He says gravity is not a fundamental force. That it instead emerged from interactions in the information that fills the universe. He claims that information, not some fundamental force, is the basic building mechanism of the universe. And his way of thinking would even do away with the need for a dark matter hypothesis."

Behind me, Dibby says, "Yes! That's what I've been waiting to hear! Gravitation is not a fundamental force. It is an emergent force that came from the sudden flood of dimensional information that shocked the cosmegg enough to split it into mirror-image twins. For your upper bubble, that happened at the quantum scale as matter and energy emerged into dimensionality.

"But there's one more big piece of information you need to notice about gravitation. It has two poles, one in each bubble. Currently, your physics in this upper bubble says gravity is the only monopolar force. Then the physicists wonder, 'How does gravity manage to hold the universe together?'

"They can't figure out how gravity works. For instance, Edwin Hubble's redshift data led them to assume the universe is expanding constantly. On the other hand, Fritz Zwicky showed there's not enough matter to make gravitation work right. The universe doesn't weigh enough to hang together gravitationally. And yet it does. So how does gravity do its magic trick?

"Obviously, some unknown factor must be holding the universe together. In desperation, physicists began to hypothesize various unseen factors. One is *dark matter*. Some say a halo of dark matter must somehow exist around objects, a mysterious kind of dark matter that emits no light or heat.

"They hypothesize that dark matter is invisible except for adding enough weight to make gravitation work right. For instance, they say your local Milky Way galaxy must have a dark matter halo around it that adds enough extra weight to hold it together gravitationally." Some of us nod. Others look puzzled.

Dibby asks, "Do any of you recall touring the *Relativity Reach?*"

I speak up. "Sure. It was interesting."

"Yet back there, you walked right past a better way to explain why the universe holds together. You all did. You missed spotting a great solution in the *Relativity Reach*...just by using a different version of Einstein's cosmological constant."

I whisper to you, "Cosmological constant? Did we miss that?" You nod.

The big bird nods, too. "Yes, you all did. I'll quote from the *Sloan Digital Sky Survey* map of the universe: 'When Einstein developed his theory of gravity in the General Theory of Relativity, he thought he ran into the same problem that Newton did: his equations said the universe should be either expanding or collapsing, yet he assumed the universe was static. Einstein's original solution contained a constant term, called the cosmological constant, which canceled the effects of gravity on very large scales, and led to a static universe."

"Wow," I exclaim. "You've got that website down to memory, Mr....uh, Bird?"

"Oh, just call me D.B. It's short for Double Bubble. And yes, my unified mind in the lower bubble remembers everything. Everywhere. In all time."

Many of us say, "Wow." "Huh?" "Lordy!" "Hot damn!" "Shit!" "No way!"

Someone whispers, "Did that bird say to call it Dibby?"

"It's D.B. But sure, you can call me Dibby. I remember all the past. I remember what's ahead, too, but it's still in fuzzy fractal outline. Anyway, about Einstein's constant. Edwin Hubble found strong data indicating that the universe expands constantly. So Einstein, embarrassed, abandoned his cosmological constant that kept the universe's size static. George Gamov even claimed that Einstein groaned to him, 'The cosmological constant idea is my biggest blunder!'

"Too bad that Einstein retreated from his concept, though. Later on, physicists started wondering if his cosmological constant actually proposed something right and vital about how gravity works. Counterbalance!"

"Yes, counterbalance!" Rodger Thompson says, nodding. "Many of us think the known properties of dark energy are consistent with a cosmological constant. The two appear to be mathematically identical in general relativity."

Dibby says, "Humph! Ed Witten and Petr Horava got closer to the truth. Their model led Mike Duff to say, 'One can speculate that all visible matter in our universe lies on one wall, whereas the dark matter believed to account for the invisible mass in the universe resides in a parallel universe on the other wall.'

"I concur with that idea of parallel universes," Dibby says. "In this Double Bubble universe—in me—a gravitational pole exists in each of my two conjoined, mirror-twin bubbles. This upper bubble that you people live in was polarized to get matter riding in 3D space and energy riding in ½D time.

"But my reversing-mirror twin bubble down below the membrane interface? It was polarized to get antimatter riding in ½D space...whose gigantic spatial pressure crush-converted it into speedy tachyon energy that powers my unified brain. It's full of complex energy patterns spread throughout 3D time. My two primal carriers, space and time, have two primal loads, matter and energy. Their reciprocal scaling in the conjoined bubbles even lets the two bubbles sit inside each other, providing the *push-me/pull-you* that makes gravitation work."

Someone protests, "But science doesn't recognize mirror-twin bubbles...."

Dibby interrupts, "...nor does science even recognize that this Double Bubble universe is a hologram. And it's not just a theoretical hologram made by bouncing two related theories off each other. Nor is it a second-order hologram made by light, like in *Star Wars,* where holographic Princess Leia delivers a message to Obi-Wan Kenobi. No, me, I'm a first-order hologram."

You say, "In *Star Wars,* that Princess Leia hologram looks tiny but real. She moves around." Then you take a credit card out of your wallet and view it thoughtfully. "But this 3D image on my card? It doesn't look real."

Dibby says, "No. It's just an image embossed on foil. That is a *reflection hologram.* It's lit from the front to reflect light into your eyes. But the 3D Princess Leia? That's a *transmission hologram.* It's lit from the rear. So you can observe it from several angles, and its photons make interference patterns in the light going through the hologram to reach your eyes. It looks 3D-real."

I pipe up, "Light is photons. Particle-waves of electromagnetic force."

"Yes," Dibby nods. "Any hologram that your technology makes is always just light. You produce a recorded, secondary image of the real thing or an image existing somewhere else. It may look real, but a light image isn't the real object.

"Light, photons, electromagnetism…they're aspects of a secondary force that appears after gravitation. But this universe is not embossed on a credit card. Nor is it a projection singing onstage. Not a holographic headset image. Not an old-timey stereoscope view, where your eyes and brain reassemble two different 2D views into a 3D picture. This universe is not just light fooling your eyes.

"Instead, all of your senses work to reassemble the hologram of this universe. It is maintained by information constantly generating me at the membrane interface. Although I'm made of polarized pulsing, to you, I'm sort of like the constant generation of a movie projector's beam of light. To you, this continuous projection of my holographic universal being creates the effect of space constantly expanding outward, although it actually does not."

People are exclaiming, "Space doesn't expand?" "No, it does!" "How dare you!"

Someone else shouts even louder, "Does it make us second-order real?"

You grin at me wryly and mumble, "Who ordered that?"

Dibby exclaims, "No! You are not just photons. Nor am I. This hologram that you live in, that each of you is a part of…it's my Double Bubble body and mind! Me, Dibby! You're sorta like my organs, only smaller, so maybe you're more like bacteria. But no, you are *not* made of mere interference patterns in photons."

The shouter asks, "So we're not just bent light?"

Dibby's wings spread to embrace everyone, everything. "No! You come from split gravitation that appeared in your bubble when matter and energy emerged. Face it, you're all part of me, and I sprang from the cosmegg's pulses of sheer being and nonbeing that sophisticated into polarized pulses on many levels.

"Polarized pulsing in the membrane interface at my middle keeps projecting 3D space upward and 3D time downward in the myriad hourglass cells that merge holographically into my two huge bubbles. They are conjoined twins with mirror-reversed features of polarized space and time holding mirror-reversed features of polarized matter and energy. My two bubbles contain everything in this universe, even each other, at opposite ends of the sizing spectrum."

"Right now, at my membrane interface, that pulsing still operates. I'm still just nothing but the potential discovered in the number 0. More exactly, I'm the linear, binary potential in nothing and something viewed as 0 and 1. I'm also the networking, analog potential in the 0 of nothing polarized into two somethings: -1 and +1...which can be subdivided into many events relating to each other."

You whisper, grinning at me, "So we're all a zero and also everything?"

Someone else exclaims, "This blows my mind! Are you God?"

"Oh no! I'm just the Double Bubble universe. One universe of many. But when I recognized my potential to make something of myself, I struggled to live, not die. I elaborated my pulses of being and nonbeing into hourglass cells of space and time, with the leftover pulses recycling into matter and energy. Some of it emerged in your bubble at the quantum scale as particle-waves. As you."

The shouter asks, "Doesn't that mean we're second-order?" Particle-waves?

"No, it means you're not just photons! You're as real as I am. Look, I'll tell you what's not real! Dark matter and dark energy! They are mere chimeras. They're like the fabled unicorn that ancient explorers saw on a far horizon... and turned out to be a real oryx antelope standing in silhouette on a hill.

"Dark matter, dark energy, time travel...those surmises come from not seeing my other pole of gravitation in the mirror-twin bubble. From not realizing I'm smart, due to antimatter that crush-converted into tachyonic energy powering my big brain. From not seeing the arrows of time and space constantly 8-looping in a tensor network across both my bubbles, repolarizing as they switch bubbles, with all of it balancing out my Double Bubble body."

A babble of voices: "Hmm." "Okay." "No way!" "Maybe it's worth considering."

Dibby gestures at itself. "My simple but profound symmetry can explain the mysteries of dark matter, dark energy, lost antimatter, a lost gravity pole, and so on. In short, my Double Bubble hologram operates much like a version of Einstein's cosmological constant, busy balancing out my information's math. And I'll show you how it balances. Come on out to see my nest in the dp-tree."

7. The Double Bubble Nest

On the way outside, Dibby says, "My dp-tree can help your science make space and time symmetrical by just adding another footprint to that solitary, lop-sided footprint you saw on display in the Relativity Reach."

Dibby chortles and stamps on both feet, which then rotate sideways so that he stands duck-footed. "Why don't you give this big, old, dirty bird another foot to stand on? Act like good detectives! Find my other foot!"

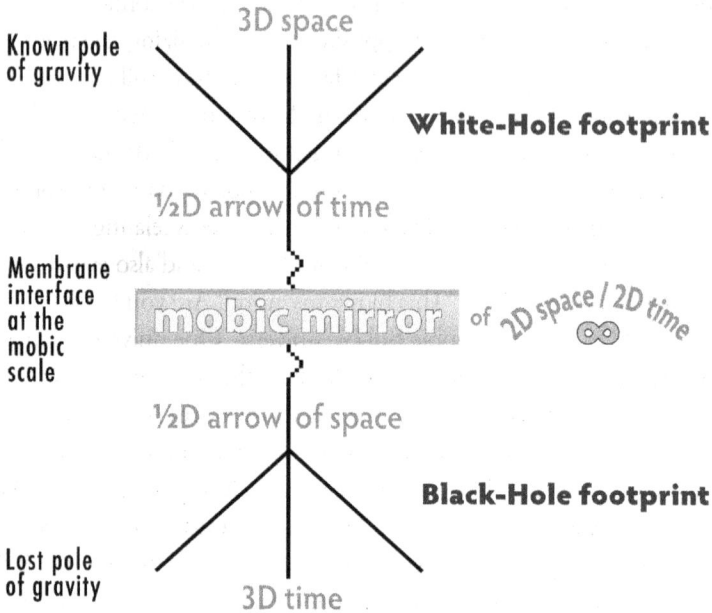

The Double Bubble's space-time footprints

"Look, this diagram even shows my other foot!" Dibby shakes each foot in turn. Wiggle: "My upper bubble has a trident of 3D space and ½D time, plus the known pole of gravity." Waggle: "My lower bubble has a trident of 3D time and ½D space, along with what your science calls the lost pole of gravity.

"I can maintain gravitational symmetry in this stance, plus I have total parity between my space and time dimensions, including charge reversal. My balanced stance provides a symmetrical Double Bubble signature. And all of it is completely in keeping with your laws of physics. It is certainly a more rational approach than the one-legged, crippled stance your current science gives me."

Dibby wiggles a toe. "Says my big TOE." We laugh, some of us uneasily.

8. Space carries matter; time carries energy

Dibby says, "In my body, space carries matter, and time carries energy. Up above the quantum scale where you live"—a sweeping gesture at us—"3D space has enough leeway to let original matter expand into pop-out molecules…and in your bubble have by now evolved into lots of tiny, diverse bodies and minds.

"But my lower bubble's ½D space could not let its original antimatter expand. Luckily, the only way that antimatter could escape ½D space's horrible pressure was by crush-converting into tachyon energy that escaped space to ride in 3D time. Its power evolved my speedy unified brain."

Someone says, "What's all this about ½D time, ½D space? Our time is 1D!"

"Not really. It is only ½D. It goes in only one direction: ahead. If your time were 1D, it would extend both forward and backward. And your omnipresent ½D time vector is just the upper half of my tensor network that's constantly 8-looping across both bubbles, moving ahead and switching polarity whenever it crosses the membrane interface of my porous passageway between bubbles."

Someone frowns, murmurs in a doubtful tone, "Tensor network of arrows?"

"Okay, get this. The mobic scale projects many hourglass cells. Each has 3D space above and 3D time below. Each cell also holds three 8-loops. When the hourglass cells merged holographically into this Double Bubble universe's two conjoined bubbles, the 8-loops formed a tensor network running on a continuous trajectory across both bubbles. Each 8-loop repolarizes from space to time, or vice versa, upon crossing the mobic-scale interface between bubbles.

"You folks up here have pop-out molecular bodies residing in the 3D space lattice, so you'll experience only the upper half of those circulating 8-loops. You'll experience it as time's arrow moving forward on a one-way, ½D path, creating an omnipresent *now*. In this bubble, it's always *now*. You can't escape *now*.

"And many of those time arrows carry energy loads coiled on them. When an arrow in your spacetime lattice reaches the right scale for its energy load to manifest to you humans, you experience it as a wavelength.

"But my lower bubble's 3D time lattice has the lower half of all those 8-loops. They create its constant, omnipresent, one-way, ½D arrow of space moving forward. In that bubble, it's always *here*. Everywhere. There's no *over there* there."

Some of us murmur in puzzlement, awe, protests, annoyance, or queries.

"Hmm, I see you're confused. Even resistant, some of you. Okay, climb up this dp-tree trunk and look into the owl-hole. And wait, let me set my owl-hole TV to the right thought experiment. Okay, now look!"

In the owl hole, we watch a 1940s John Wheeler talking to bongo-playing Richard Feynman. "Dick, suppose that time is a line tracking a single electron in space. The electron's timeline gets all tangled up, turning into a big, messy ball."

Suddenly a big, tangled ball of timeline sits beside Wheeler. He pulls out a big scimitar and whacks the big, tangled timeline ball, slicing it right in half!

"Voila!" Wheeler exclaims. "Dick, I've just cut the electron path at multiple points. Now, if I stand at a cut on the electron path and look in one direction, I see it as an electron moving forward in time. Correct? But if I look the other way, I see it as a positron going backward in time. Reversing its path on the timeline turns the particle we call an electron into its antiparticle, a positron!"

"How whimsical," Feynman says, drumming. He nods slowly and muses, "Yes, John, it's brilliant! So maybe the identity of matter or antimatter depends upon where we stand upon the timeline to observe it. Maybe our human point of view always stays tucked behind the cut somehow, and thus we interpret the points to mean our universe is made of matter, not antimatter. "

Dibby says, "Wheeler's idea is a rather tangled way to describe the reversing-mirror effect of my Double Bubble body. Above, my 3D space and ½D time hold tardyonic particle-waves of mattergy, what you folks call normal stuff. But below, my 3D time and ½D space hold tachyonic particle-waves of antimattergy.

"You might consider my Double Bubble body like a reversible glove." Dibby morphs to show just the upper bubble. "My white-hole bubble has matter in 3D space and normal energy in ½D time. Up here, your myriad separate mini-minds exist in tiny molecular bodies that move in 3D space over ½D time."

Dibby flips inside-out. "But this black-hole bubble holds my unified brain that's spread throughout 3D time. Speedy tachyons power its shifting energy patterns. Meanwhile, this black-hole bubble's ½D space holds tachyon particles with imaginary mass. Imaginary in the number sense, that is, as negative mass."

Popping back into a Double Bubble shape, Dibby says, "And how do you notice my mind? You say it's Mother Nature. Evolution. Even God.

Both kinds of mind—your many, tiny minds in 3D space and my single, unified mind in 3D time—we even communicate through the mobic scale in dreams, meditation, trance, insights out of the blue, and other altered states.

"For instance, dreams come from what you call the collective unconscious. They're short, personalized messages from me, but nowadays, when you wake up into ordinary consciousness and your big, fancy egos come online...at least, in the *homo sapiens* species.... you mostly just call the dreams weird nonsense."

I'm nodding vigorously now because it corroborates something I've already been pondering. We humans already do recognize the mysterious property of shared mind operating in many insects: ants, termites, and bees. Some other species also exhibit hive-mind or group-mind traits. So mind must be far deeper, wider, and odder than we like to think. Other, less "evolved" species might even tap almost continuously into the guidance of universal mind.

After we climb down from the dp-tree, I say to you, "Okay, maybe physics really does need a big TOE that will balance out that single footprint of a currently lame bird. It's standing unsteady due to all the physics issues that leave it lame and wonky. Maybe the Double Bubble's two space-time footprints would fix it. Give it another leg to stand on."

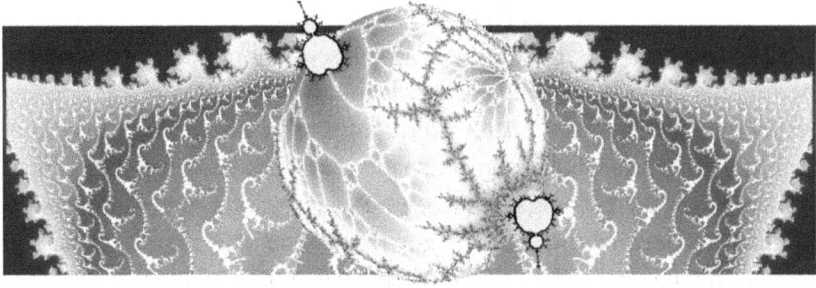

Chapter 4: The Global Shakedown Cruise

1. We voyage on a global shakedown cruise

Currently, the globe is starting to rise above a myopic loyalty to just one nation, language, culture, cuisine, or mindset. The new, worldwide gestalt will be sorting out what works best amongst all nations, cultures, societies, cuisines, games, hobbies, habits, even religions. We're all voyaging together on a global shakedown cruise toward a new sense of oneness in humanity living on Earth.

I participate in that cruise. Born in the USA, I've also lived for years in Turkey, France, Switzerland, and China, along with traveling widely. As I tuned into the unfamiliar setting in each new locale, I tried to absorb its ambiance and discover new things that interested, amazed, shocked, and delighted me.

When I went to live in Guangzhou as a university teacher, it was startling to enter such a sprawling, noisy, messy, lively city, especially right after living for 5 years in tiny, tidy, manicured, scheduled Switzerland. The sudden switch to living in China kept tweaking new adjustments of attitude in me.

I found that my students at Jinan University were a real pleasure to get to know. Along with the expected local and regional students, the children of many diplomats and overseas workers also studied there. Generally speaking, they were unjaded yet refined, and so eager to learn. They were adept at juggling hierarchies of abstract thought and symbolic meaning.

They also tended to voice their opinions rather diplomatically...not just to me as a foreigner, but also usually to each other. I soon realized they had a more holistic mindset than my former students back at the University of Texas, many of whom lived for frats and sororities, sports, drinking, and hooking up.

On jaunts with Chinese teachers and grad students around the city's sights on weekends, we talked poetry and drank glorious pineapple beer from long-necked, brown bottles. I decided China holds supreme distinction whenever it follows its sure instinct for the good, new thing while also honoring its ancient culture, the diverse cuisine, the surprising architecture, and the traditions of expert handcraft skills that became translated into technological dexterity.

2. The holistic mindset

The traditions of China fostered a philosophical mindset that ventured beyond the limits of linear logic. Such a mindset did not consider ongoing reality merely in terms of matter's linear units to be summed up. Instead, it looked at events more holistically as dynamic timing patterns in emergent flow.

Long ago, the I Ching's scholars were scientists of a sort, working with numbers to describe the dynamic flow of patterns that shape events. Like modern-day chaos theorists, they understood that the timing of energy in the spacing of matter generates patterns in events. Those patterns have general characteristics that can recur with many variations of their specific details. It happens on scales large and small, and all of it puts fractal patterns into play.

Kristofer Schipper and Wang Hsiu-huei said in *Progressive and Regressive Time Cycles in Taoist Ritual,* "In Chinese thought, the universe is apprehended as an infinity of nesting time cycles that, because of their formal correspondences, may be manipulated as though they were interchangeable...."

The ancient mindset revealed in the I Ching emphasized dynamic patterns in events and their connective relationships. Such a mindset saw events as a purposeful flow on a gradient, moving all things toward the sea of meaning. It is the watercourse way of the Tao.

A modern instance of that mindset appears in a 1992 poster that I bought while on a biking trip to a village near Guangzhou. A street vendor sold it to me. He said the poster was printed to celebrate the upcoming Spring Festival. I still have it. Its wide red border frames a corpulent Buddha, a trio of Confucian gods of benign plenty, and even a golden bat flying above them to symbolize a Taoist ability to see things hidden in the dark.

Flowing down one side of this rosy, jolly scene are Chinese characters for...

THE LUCKY STAR'S RAYS TO EARTH BRING 100 WELCOME HAPPINESSES.
Down the other side are the characters for...
THE BOUNTEOUS SPRING BRINGS 1000 BLESSINGS.

This poster uses relational, analog numbers to indicate intangible wealth. Its numbers, 100 and 1000, are not a progression made by adding units of 10, but rather, by raising 10 in exponential leaps, as in $10^2 = 100$, and $10^3 = 1000$. It bespeaks the holistic, relational, analog mindset of traditional China rather than a modern Western mindset counting out material wealth in discrete, additive chunks. You can see that old poster on YouTube. It appears in the background of two interviews that Paul O'Brien did with me in 1997.

After that biking trip, back in my university flat with that poster mounted on my living room wall, I watched a Chinese professor sit on my couch and gaze at the poster in disdain. He informed me that it portrayed a wholesome, robust perfection that would appeal only to a peasant audience.

And to me, I thought. The poster had no urbane, intellectual edginess of "the modern face of China" that the professor hoped I would notice. Instead, it revealed an older Chinese gestalt where the symbols of wisdom in three different religions could share their poster space quite companionably. It embraced Confucianism, Buddhism, and Taoism all at once…in the *both-and* style of syncretism that has long been typical of Eastern religions.

In the West, one cannot be Jewish, Christian, and Muslim all at once. Each of those spare, desert-born religions absolutely demands sole allegiance in an *either-or* stance. Meanwhile, the Eastern religions tend toward a syncretic *both-and* inclusiveness. In general, the West conquers; the East absorbs.

3. China opted for coherent stability

Although early China appreciated holism, it was also well ahead of the West in science and technology. Emperor Yu (c.2200-2100 BCE) began dynastic rule and became famous for devising an ingenious flood control system. Marco Polo 3,500 years later revealed many marvels of Chinese culture to medieval Europe.

Western scholars admired 4 early Chinese inventions—the compass, paper, printing, and gunpowder. Those 4 advances were key to China's national unity because they allowed printed edicts to be spread in a shared written language across the land and then be enforced. Early on, central rulers could impose a strong national identity on Chinese culture, which tended to unify it.

In *Studies in Asian Frontier History*, Owen Lattimore said, "Ever since the first emperor of the Qin Dynasty conquered six other states and assumed absolute authority [221 BCE], the form of a unified and centralized feudal universal state has always predominated in the social structure of China."

In China, scientific and technical development rose in a continuous gentle curve from about 1500 BCE to about 1600 CE. That's 3,100 years. It's a long time span for a country to have such slow, steady growth. China lived in coherent stability for so long because its feudal system differed radically from Europe's highly fragmented and competitive groups who pitted diverse rulers and values against each other. China instead emphasized far greater unity of government, language, money, culture, and standardized weights and measures.

We might imagine a similar timespan of coherent stability for the West if Imperial Rome had never fallen…if instead, the Caesars had extended a mighty umbrella of Latin language, government, inventions, war machines, and unified culture over a vast Roman empire fostering scientific and technical development on up to now…and then onward for 1,000 more years.

4. Outstanding Chinese science gradually fell behind

Early Chinese inventions were amazingly diverse, practical, and forward-looking. For instance, in the Song dynasty (960 to 1279 CE), Chinese textile machinery far surpassed that of the West. According to the ancient *Shihuozhi* record of Song dynasty history, the Chinese had by then invented animal- and water-powered spinning machines with as many as 32 spindles. They were capable of producing 30 to 50 times more than manual spinning wheels produced. The silk submitted by various districts for taxes was 3.41 million bolts. The West only found such technology during the Industrial Revolution.

Then why did science and technology in China, once so successful, gradually fall far behind Western science and technology? It's called the *Great Divergence,* and scholars offer a wide range of theories to explain it. In my opinion, a ritualizing tradition in China began to devalue the logical chain of systematic query and application that is so vital to scientific theory and testing. Instead, a vast lexicon of symbolic flattery gradually grew up around the emperors and their kowtowing courts, reinforcing customs that fostered a stale, centralized bureaucracy dictating all policies, finances, education, morals, and manners.

Custom quashed query and reinforced such stability that public life gradually petrified into empty rituals rendering the rulers, their subjects, and the economy hidebound. China feared change, viewing it as synonymous with chaos. But by resisting change, that profound, ancient culture became so equilibrated that for about 3,200 years, nothing could knock it out of its ultra-stable entrainment and move it onto a faster track of cultural evolution.

China finally began to be confronted with Western trade and imperialism, shocking it into a modernization that was not cultural evolution, but rather, cultural revolution. The Chinese Revolution of 1911 sought to address the nation's ritualized stagnation by bringing down its last young emperor, Puyi, but without finding a truly unifying leader to replace him.

Japan's invasion of China in 1937 fueled anger and unrest that sparked another Chinese cultural revolution in 1949. Its newly communist society championed a strong leader, Mao Zedong, who offered a tweaked Marxism. Dialectical materialism rejected the feudal stagnation of customary ritual and promised a rosy-Red dawning of material plenty to be shared equally among all.

Chinese communism repudiated the "over-idealized past of decadent aristocrats and dry scholars." By 1966, many of the young sprang up from the masses as the Red Guard, eager to enforce the Cultural Revolution. Cadres of paramilitary youths destroyed much old art and architecture, as well as punishing, shaming, or even killing those whom they deemed "anti-revolutionary."

China's bizarrely autocratic communism arose, with a small, privileged, wealthy class of the elite who ruled by keeping the masses obedient to an idealized Chairman Mao. The privileged, centralized, one-party command tried to maintain economic competency along with a group-think message.

When I taught at Guangzhou University during 1991-92, two years after the Tiananmen Square massacre, students told me they were compelled to write essays on the dangers of chaos for a year. Chinese faculty said they'd attended mandatory "political study sessions" meant to "turn around people's thinking."

On a trip to Hong Kong, I looked up old articles to corroborate the stories I'd heard. A *Newsweek* report said a Chinese university dean quoted the exact words of many official editorials when he declared, "We want to make sure everyone's thinking is unified." Other articles said officials told people to betray family members who'd deviated from the party line, and many complied. Such insistence on a submissive populace holds the echo of an earlier, imperial China.

I talked with one Guangzhou state leader who, after 20 minutes of evasion, finally boasted that he'd sat in a secure party enclave in Beijing watching on private TV the 1989 student massacre. He saw many students shot in Tiananmen Square. That 1989 student uprising was labeled "counter-revolutionary." So the word *revolution* had become ritualized in a way that canceled out its meaning.

While living and teaching in China, I eventually realized it was no wonder that China went communist after such a long, ritualized, feudal past. Steeped for millennia in the group-think of a feudal collective at the national level, the masses found communism somehow in keeping with their long tradition favoring a vast, stabilized connectivity with strong central guidance.

At first, communist China marched stolidly into the future on Red shoes borrowed from Russia, but when that Marxist dream of a sturdy peasant solidarity grounded in communal plenty did not prosper enough, China began to craft its own Red silken slippers with a custom political fit.

China absorbed the West's technological capitalism and adapted it to the values of group identity and group sharing. It also re-evaluated and reinterpreted its old, deep scholarship and prudent wisdom into modern innovations, achieving a synthesis that carried it far and fast. By attracting foreign corporations, tourists, and scholars, it added more yeast to the mix. In the 21st century, China still rises, with an increasing number of names that make important scientific discoveries.

5. Beyond nationalism to globalism to holism

This TOE says the Double Bubble universe's matter and energy have connective relationships in space and time that let the lower bubble and upper bubble communicate instantly with each other. How does it happen?

First, consider a scientifically proven instance of that communication. Experiments in *quantum entanglement* have conclusively shown that two particle-waves can interact, get separated, and yet still stay in relationship at a distance. Actions performed on one particle can instantly affect the other particle. Einstein called this event "spooky action at a distance." He said it only looked like communication because quantum mechanics not yet completely formulated.

Unfortunately, the word *entanglement* makes quantum events sound haphazardly tangled, like cooked spaghetti tossed on a plate. It is an English translation of what Austrian physicist Erwin Schrodinger first described as *verschrankung*. That word might better be translated as *enfolding* or *clasping* because casual messiness is not the norm in your body, nor is it true in the universe. You would not call your body's workings entangled. Instead of being entangled, the universe's events enfold in layers of connective resonance that David Bohm called the implicate order.

Fortunately, we can unfold nature's implicate order by using scientific tools and methods to reveal some aspects of its universal structure...which turns Schrodinger's discovery of communication between two long-distance particles into an instance of Bohm's larger view that a deeper-level ordering principle exists in the universe. This TOE says it occurs at the mobic scale where polarized pulses operate in the membrane interface between both bubbles to generate dimensionality and set mattergy into the chute of possibility. The pores of that interface continually project the hourglass cells merging holographically to form both bubbles by bringing into existence something out of nothing.

Those polarized pulses generate complexity on many scales in both bubbles, with all of it finely combed into dynamic continuity. We have atoms in organisms on planets in solar systems in galaxies in clusters of super-clusters throughout our upper bubble.

Fractal patterning shapes the four primals of space, time, matter, and energy in ordered layers as delicate as a mille-feuille pastry, as neat as a baby's downy head, as gleaming as a Lorenz attractor's gossamer lines spun by analinear numbers. At every level of reality, from the physical to the philosophical, our resonant, relational universe holds fractal co-chaos patterning. At each moment, it communicates across all sorts of boundaries. One can learn to participate in it, if the ego is willing to bend and become enfolded.

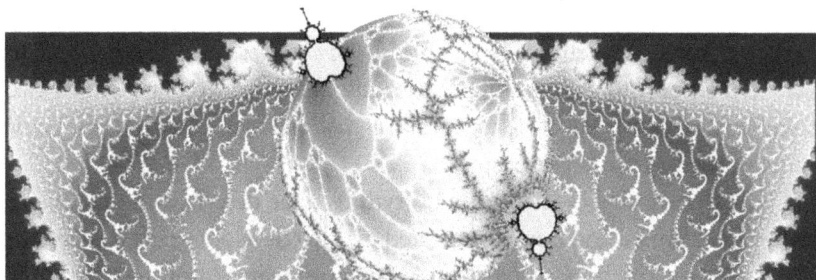

Chapter 5: The Two-Sided Treasure

1. The universe makes a point

In this chapter, we obtain tickets to ride a dimensional roller coaster exploring the Double Bubble's dimensional structure of space and time.

The dimensional roller coaster on the spacetime fabric

We jump aboard the dimensional roller coaster, but our car does not move. It can't go until dimensionality emerges from nothing…when nothing in the sea of possibilities realizes itself as something. What pings its identity? Mind. An awareness of itself as nothing…and the notion that nothing may be represented by a number: 0. This 0 is information.

Could mind be the impetus for generating information that develops this whole conscious universe? Is such a thing even possible? Some physicists consider it possible. Back in 1974, Jack Sarfatti said in *Implications of Meta-Physics for Psychoenergetic Systems*, "The full meaning of quantum theory is still in the stage of being born. In my opinion, the quantum principle involves mind in an essential way along the lines suggested by Parmenides, Bishop

Berkeley, Jeans, Whitehead, et al." And in *Space-Time and Beyond,* Sarfatti said,"…I suspect that general relativity and quantum theory are simply two complementary aspects of a deeper theory that will involve a kind of cosmic consciousness as the key concept."

This series offers a Theory of Everything that describes the conscious universe we live in, and how it generated itself by developing a fractal master code based on co-chaos patterning. A dawning awareness of nothing as the number 0 jump-started the creative potential hidden within its 0-number identity. This 0-point of number identity also has zero dimensionality.

○ The 0-point

The 0-point of dimensionality (greatly enlarged here for visibility) winks in and out of existence on the sea of possibility that is tossing it and other would-be universes. It pulses randomly at first in fitful beats, searching for the simplest way to exist. The cosmegg's 0 claims its identity simply by pulsing into being and out again, marking two states: being and nonbeing.

On finding the mobic scale, those recurring beats grow more regular in their *on-off* pulsing…sort of like a neon sign strobing on and off. Or like a drum beating. Or like the breath that thrusts from my lungs to become visible in the freezing air as I run. It vanishes. Then another breath appears. Vanishes. Or maybe it is like a heartbeat that marks the indeterminacy of life and death itself.

As the cosmegg's pulsing point becomes regular, its *on*-beats hammer at the inchoate potential still locked within its 0-number identity. Each pulse is anchoring and reaffirming the cosmegg's identity, but it also pings all its unrealized potential, its *massa confusa* of possibility. The 4 primals, the 4 forces, and indeed, all possibilities hide within 0, awaiting release into manifestation.

This regular, systematic *off-on* pulsing sets up a model or archetype for everything that will come later. Heartbeats. Births and deaths. Pulsars. Each beat reaffirms that at the mobic scale, this point of identity has found the right boundary to strive for success. The cosmegg pulses constantly in and out of being, exploring what works here and what doesn't, and it eventually evolves enough range of expression to generate this Double Bubble universe.

A particle physicist might assume this 0D point is striking up the band for early quantum mechanics, which considered particles to be point-like. Later on, it modified that idea to consider *some* particles point-like. Such a physicist might assume that I'm describing the dimensionless point hypothesized by quantum mechanics for a particle-wave emerging and disappearing again in the haze of quantum indeterminacy.

But no, this TOE says the pulsing point of sheer being vs. nonbeing isn't at the tiny quantum scale where particle-waves emerge. No, it is much farther downscale. It is at the ultra-tiny mobic scale where space and time will eventually emerge. So calling this point of identity a particle-wave is not "on point."

Simplex math would call this dimensionless point a 0-simplex, meaning it has zero spatial dimension. Thus, by that logic, a 1D line is a 1-simplex, a 2D triangle is a 2-simplex, and a 3D tetrahedron a 3-simplex.

Simplex geometry can postulate even more dimensions of space than current science and everyday experience are able to observe in our universe, so it can generalize the concept of a triangle or a tetrahedron to an arbitrary number of dimensions…but those dimensions are always in *space*, not *time*.

Simplex geometry describes space dimensions, but this TOE also describes time dimensions, so we'll call our variation *co*-simplex geometry. Thus, instead of saying the point of cosmegg identity is a 0-simplex, we call it a *0-co-simplex*— meaning it has 0 space dimensions and 0 time dimensions. This co-simplex geometry will generalize the concept of a triangle or a tetrahedron at the mobic scale to a *flexible but not arbitrary number* of dimensions in both *space* and *time*.

To foreshadow the cosmegg's potential, we'll set the *on-off* beat of this pulsing point inside a *potential mobic band* whose polarizing twist may, can, must develop its repolarizing dynamic for the universe to succeed.

A pulsing point of 0-1 identity sets a 0DD point on a potential mobic twist

2. What is a 1DD line?

Another *on*-beat of existence occurs. This new *on*-beat occurs close by, but both elsewhere and elsewhen, so the cosmegg is now iterating two different yet related conditions of polarized existence at that ultra-tiny scale. A *line* of polarized tension stretches between the two recurring *on*-beats at the mobic scale. They establish two poles of dimensionality…one of *space*, the other of *time*. We'll call this line the *1-co-simplex* of space and time, or for short, space-time.

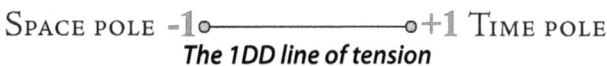

SPACE POLE -1○————————○+1 TIME POLE

The 1DD line of tension

Both space and time dimensionality have now emerged from the *massa confusa* of possibility. Both poles of this line are part of the same family— dimensionality—yet their differences also separate them. This line of tension tests the limits for dimensionality at this minuscule scale, while it also affirms that each pole can exist here as its polarized opposite other.

As the two ultra-tiny, pulsing beats anchor their separate poles of space and time, by that very act, they also anchor permanently the identity of both poles existing in dimensional relationship. A foundational paradox has originated here at the ultra-tiny mobic scale. Space and time need each other to exist. Hence, each pole's identity must refer to its Mysterious Opposite Other.

This premise of polarized identity is what primes all storied romance to seek the Mysterious Other, and in fact, all attraction and repulsion in the universe. It triggers the core paradox of all conditions, of life in the midst of death, of light needing shadow to exist, and so on. It even determines the shape and contents of our Double Bubble universe. Everyday examples of this principle operating in our own bubble are this globe's North and South magnetic poles, procreation between two sexes, a car battery with two terminals, mechanical cranes that can magnetically lift and drop huge loads of metal. Opposites attract.

And like knows like. For instance, molecular polarity is a very important factor in solvents. Like dissolves like. Polar substances dissolve in polar solvents, but non-polar substances dissolve in non-polar solvents. This polarity principle so embedded in events can hatch some weird paradoxes. Take this one: the underworld Greek *g-o-d* Hades had a fierce *d-o-g*, Cerberus, who welcomed the dead crossing the River Styx into Hades, but would let no one leave. Yet in his final labor, merely human Hercules resolved that god-dog polarity.

Below, the 1DD space-time line sits in the mobic twist of a *potential loop*. Its two polarized ends are space and time, symbolized by -1 and +1. Its poles have an analog, relational quality that acts quite different from that other number mode of *off-on* pulsing in binary units of 0-1. Together, both modes of number—linear, sequential, binary 0 and 1, and analog, relational, polarized -1 and +1—give the cosmegg two different ways to get past 0's nothing identity, and both modes working together can fast-track its analinear dimensional development.

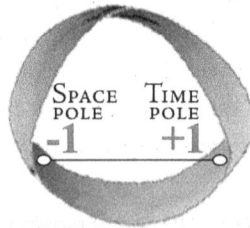

SPACE
POLE
-1

TIME
POLE
+1

2 pulsing points set a 1DD line of space & time in the potential mobic twist

With the 1DD line, we start moving on the roller-coaster track of dimensionality. Out beside the track, we notice a loud and proud String

Theory jazz band strutting and improvising riffs on string theory's claim that it is lines, not points, that create particle-waves at the quantum scale. But hey, the cosmegg isn't even that big yet! We're still way down at the mobic scale!

3. Viewing the p-tree's first fork of dimensionality

Our foray into dimensionality will develop levels of polarized bifurcation, and the easiest way to depict this is with a *polarized bifurcation tree (p-tree)*. The blinking cosmegg's *on*-point of existence sprouts a trunk that forks into two polarized branches. This first forking records the cosmegg's evolution from a mere *off-on*, 0-1 blip of existence and nonexistence into two polarized *on*-points as states of space (-1) and time ($+1$). They are the two poles of its 1DD line of tension, where the -1 pole balances its equal but opposite $+1$ pole. The stretch between space (-1) and time ($+1$) at the mobic scale is ultra-tiny.

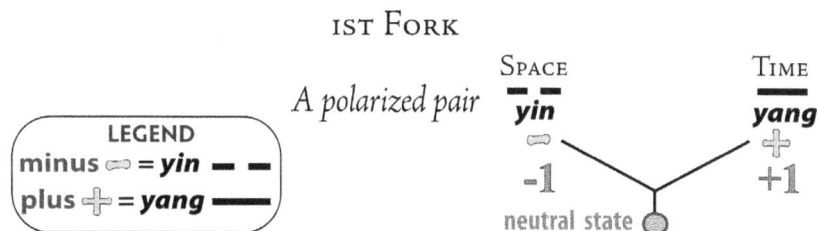

1ST FORK

A polarized pair

SPACE
yin
-1

TIME
yang
+1

neutral state

LEGEND
minus = *yin*
plus = *yang*

The 1st fork of the p-tree describes a 1DD line polarized by space & time

This infinitesimal stretch tests the scaling limit at which both poles can exist separately yet still communicate to keep the bonds of relationship going within their dimensional family of origin. Eventually, the cosmegg will create higher levels of more complex dimensionality, and via polarized forking, the p-tree will sprout both branches and roots to become a *double p-tree* (dp-tree).

Simple *minus* and *plus* are too unrefined a polarity for this task. We'll need a special shorthand to express the dp-tree's levels of increasingly polarized dimensionality. Fortunately, ancient China's I Ching math symbols can easily shorthand the needed levels of polarized dynamics. Thus, on the p-tree above, *minus* (-1) equals the broken line of yin ▬ ▬, while *plus* ($+1$) equals the solid line of yang ▬▬▬. Now we can also drop the 1s and just keep the polarities.

4. Look, it's a 2DD plane!

The cosmegg is pushing for more growth, seeking a new, higher-order dimensionality. And here it comes! A new beat occurs at an infinitesimally different location in space and time. This new point is equidistant from the other two already-established points. The tension path around them is within this ultra-tiny scale's limits, so this new third point generates *area*.

The cosmegg is no longer just an *off-on* pulse establishing its being or nonbeing as a flickering 0DD point. Nor is it just two polarized points of space and time establishing the dimensional tension of a 1DD line. With the addition of a third point, the single line of tension abruptly multiplies into *three* lines of tension linking all three points! The cosmegg holds a tension path around all three pulsing points, sketching in the area of a 2DD triangle.

You already know, of course, that an ordinary triangle in our upper bubble is a plane of 2D space, as Euclid's classical geometry defines it. Either side of that triangle exists in 2D space, so each of its faces shows a 2D space area.

However, the triangle at this ultra-tiny scale is different. Yes, it has two faces...but they are a 2D space face and a 2D time face! In other words, its 2D space face is the polarized opposite of its 2D time face.

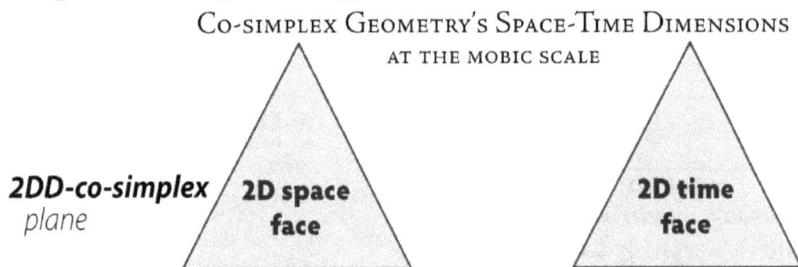

CO-SIMPLEX GEOMETRY'S SPACE-TIME DIMENSIONS
AT THE MOBIC SCALE

2DD-co-simplex / **2D space** \ / **2D time** \
plane / **face** \ / **face** \

This 2DD triangle has two faces: 2D space & 2D time

To grasp this idea, let your inner eye visualize the 2DD triangle. You can see the 2D space face. Flip it over to look at the other side. That face is polarized by 2D time. It's sort of like flipping a coin. You flip it and get either heads or tails, either the 2D space face or the 2D time face, depending on which side is up.

Think of it! By adding just one more blinking point, the cosmegg has now established 4 planar dimensions at this ultra-tiny scale, and they all work together. How does the cosmegg maintain so much dimensionality from just three points? By constantly iterating those pulses at those three locations! Its pulsing iterations can maintain the 2DD triangle, sort of like the way your iterating heartbeat maintains you.

This triangular plane faced by 2D space and 2D time is a 2-co-simplex. Note that a triangle's 3 points provide the most economical and efficient way to rise to this new, higher order of dimensionality that captures *area*. According to this TOE, the three points of a triangle must be used, not the 4 points of a square, rectangle, kite, or trapezoid. Why? Because triangular tension sits at the sweet spot of bonding capability as the closest/furthest distance for optimal result at this ultra-tiny mobic scale.

Luckily, a triangle holds 180°, while a circle holds exactly twice that number:

360°. This foreshadows the possibility that a 180° triangle will accommodate perfectly the 8-looping dynamic of a potential 720° mobic-loop that you see forming around the burgeoning dimensionality at this ultra-tiny scale.

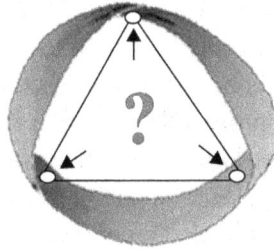

What is the polarity of the 3 points on a 2DD plane of space & time?

You may be protesting, "But wait! Hold on! How can the two poles of yin ▬ ▬ and yang ▬▬▬ fit onto a triangle's three points? Sure, it was easy enough to put two poles symbolizing space and time on a line's two ends. That seemed sort of like the North and South poles on a compass needle.

"But you cannot fit just two poles onto all three points of a triangle! Polarity won't distribute evenly on the triangle's three points. So the math of 2 poles shared around 3 points just won't go. Not evenly. Because two poles cannot manage to distribute properly onto three points. How could they?"

How? Look, I'll split this 2DD triangle to show you each face's polarized points. The space face reads as three yin poles, but the time face reads as three yang poles. All three yin poles and three yang poles make a 6-pack, a *pair of polarized triplets*...so imagine that a reversing mirror sits between both faces...

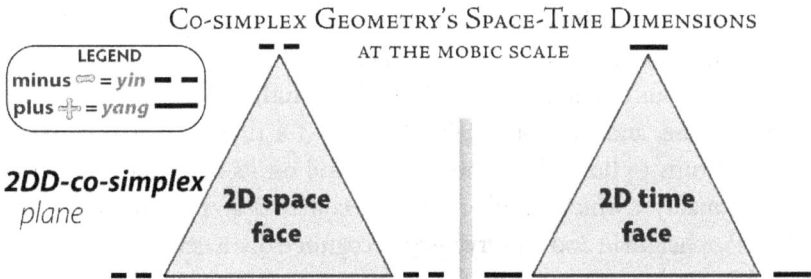

CO-SIMPLEX GEOMETRY'S SPACE-TIME DIMENSIONS
AT THE MOBIC SCALE

LEGEND
minus ⊖ = *yin* ▬ ▬
plus ⊕ = *yang* ▬▬▬

2DD-co-simplex plane

2D space face

2D time face

The two faces of the 2DD triangle have mirror-reversed polarity

...which presents us with an interesting differentiation: the 2DD triangle has a three-point origin, but it has six poles on both faces of space and time. This new, higher order of dimensionality has multiplied the yin ▬ ▬ and yang ▬▬▬ poles, sort of like the way a magnet multiplies its poles if you cut it in two.

Some physicists might assume that this plane is a membrane, M-brane, or worldsheet. But no, this two-sided triangle of 2DD space and time is far tinier. It is down at the mobic scale, not up at the larger quantum scale. The only place in the Double Bubble universe where this unique dimensionality of 2D space and 2D time can exist is at the mobic scale.

Eventually, it will exist in each mobic pore of the membrane interface conjoining both mirror-twin bubbles. But of course, all that hasn't happened yet. Only one 2DD triangle exists, but it foreshadows what will come.

Preview: the matured dp-tree will project fractal co-chaos patterns into the two polarized bubbles. They load space with matter, and they load time with energy. Within each bubble, lesser variants of the core paradigm apply.

For instance, each bubble holds one pole of gravitation. Why? Information overload split the cosmegg, blew both bubbles, put a pole of gravitation in each bubble, and established the mobic membrane interface between them. Each bubble will also develop three later-emerging forces, each with two poles. Those forces will differ per bubble due to their polarized, mirror-twin specs.

Riding on our dimensional roller-coaster, we notice that out beside the track, a marching band is high-stepping in triangular formation. Everyone is playing triangles, and the color guard waves banners proclaiming 2DD Rules!

5. The 4 stages of growing the 2DD triangle

The 2DD triangle has established its routine. We watch it continually pulsing through 4 stages that maintain and empower the cosmegg:

First—three polarized pulses sketch a tension path around a triangle's three points, and a fourth pulse takes it back home so fast that all the pulses seem just about simultaneous.

Second—Achieving that return to origin lets the cosmegg re-enact the trip around the triangle, retracing its path repeatedly, zipping around it via pulses of *being* that surmount *nonbeing*—**1-thud, 2-thud, 3-thud, 4-thud**—back to home base, and then out again. This 1-2-3-4 rhythm uses that fourth beat to return to home base and seal the deal on its new, higher order of dimensionality…which simultaneously becomes Beat 1 of the new cycle. Maria Prophetissa in 2nd century Egypt recognized this renewing momentum of 1-2-3-4 in her words: "Out of the third comes the one as the fourth."

By generating and then repeatedly renewing the 2DD triangle with its polarized faces of 2D space and 2D time, the cosmegg repeatedly confirms that the mobic scale is a reliable place to set its foundation. Each triangular trip verifies again and again that "Yes, this route is so stable, and I can do it in a flash! Sure, this triangular trip gets boring, but at least it has a plus side and a minus side."

Third—The cosmegg can keep on tracing this same old path and remain ultra-tiny. But so much inchoate potential still exists in its 0-seed of origin, unrealized, fomenting to get out. What if the cosmegg can develop even more dimensions? It is crammed with so much potential!

Fourth—Our cosmegg is the original advocate of sustainable evolution since so many other proto-universes around it are guttering out in the sea of possibility. Its 2DD triangle's 6-pack of polarized tension provides a foundation so dynamically stable that it will let the Double Bubble universe eventually grow a polarized dimensionality that is expansive, enduring, and fulfilling.

6. The wonderful weirdness of 2D space × 2D time

I admit it. By our upper bubble standards, an ultra-tiny triangle faced by 2D space and 2D time at the mobic scale seems weird. It is not the familiar Euclidian triangle that you crayoned onto a sheet of paper when you were a tot. That's because your body exists here in our upper bubble's 3D space with a one-way arrow of time, not down in the membrane interface. In fact, you live in such different dimensionality up here above the quantum scale that you could not possibly draw a true triangle faced by 2D space and 2D time.

I also admit that some people (Buckminster Fuller, for instance) have viewed the familiar Euclidian triangle…lauded for millennia, drawn on sand, papyrus, parchment, blackboard…long history!…as emblematic of an impossibly perfect, timeless, ideal triangle in the ancient Platonic sense of *ideal* versus *real*.

But now, we are contemplating a triangle that is even more exactly perfect and unique. This 2DD triangle of tension optimizes possibility at the mobic scale. Its information describes a sturdy, triangular truss born of just 3 points. And it offers the amazing bonus of 2D space and 2D time by allowing 4 dimensions to coexist at once. Note, its dynamic combines the 3's mutative *area* with the 4's reinforced inner stability. At no other scale can this ever happen.

Tension tightens around the 2DD triangle, shrink-wrapping it into the best possible fit. Simplex geometry in the upper bubble would call this tight fit around a planar triangle the *convex hull* of a *2-simplex*. But our triangle sits way down at the mobic scale. It describes both 2D space and 2D time, so it is a *2-co-simplex*. The tight fit around both of its faces is a *convex, doubled, polarized* hull…or for short, a *cdp hull*….or for fun, a *seedy pea hull*. (By the way, *cdp* is also an acronym for *continuous data protection*, and it applies here).

7. Dimensional development sets up the Rule of 4

To get the scope of what has been transpiring here, let's pause to review the process thus far—so we can then extrapolate into the future. By the way, in this series, I write numbers as digits whenever I want to emphasize their math aspect. For example, instead of two and four, I'll write them as 2 and 4.

We've seen how the original 0-1, 0-1 blink of *off*-and-*on* pulsing developed identity's habit of 2-ness. Then an *on*-beat in a slightly different location gave a new variation on 2-ness. It created a 1DD line of dimensional tension connecting space's negative pole and time's positive pole. Just one more beat in a new location in space and time described a 2DD triangle. It had 2 faces—a 2D space face and a 2D time face.

Just 4 pulses have now made 4 dimensions, and they are a polarized pair of pairs. With that, habit solidifies into stability, and stability develops a rule to guide the cosmegg's evolving dynamic. This *Rule of 4* ensures enough stability to maintain the cosmegg's burgeoning growth into the Double Bubble universe. It will let the cosmegg develop a master code that becomes a giant fractal generator for the 4 primals—space, time, matter, and energy—a polarized pair of pairs.

This provides the Double Bubble universe with 2 reliable carriers—space and time. They are polarized to carry 2 fluxing cargoes—matter and energy. Innate polarity decrees that space will carry matter and time will carry energy. As with all fractals, the basic forms (space and time) will remain recognizable, but the variable contents (matter and energy) will flux in their specific details.

The universe's 2 carrier forms, space and time, will continue to develop the master code by establishing dimensional structure via polarized pulsing. Its pulsing has already generated a 2DD triangle's 6-pack of dimensional polarities. They're a *pair of polarized triplets* operating in a co-chaos pattern. Think of them as a polarized dimensional 6-pack.

The universe's two content loads, matter and energy, will emerge at the larger quantum scale, using an aspect of the master code to establish particle-wave structure. The mattergy load in this upper bubble and the antimattergy load in that other bubble make fractal iterations of particle-waves in a changing flux of specific details. The Double Bubble universe exists by maintaining its dimensional form and evolving its mattergy contents. Like we do.

Even so far above the mobic scale, we still notice the Rule of 4 with its polarized pair of pairs. It manifests to us in ways that range from the physical to the philosophical, but most often, we see some attenuated version of it... for instance, in a compass with 4 directions. Or a torso with 2 arms and 2 legs. Or a car with 4 counter-balancing wheels. We live among many minor knockoffs that are fragmentary riffs on the foundational Rule of 4.

It happens because the universe considers any 4-some relatively stable, and whenever possible, that 4-some will act as a polarized pair of pairs. However, up at this coarser-grained resolve above the quantum scale, it may devolve into just a quartet. For instance, in our upper bubble, we notice how nature likes to manifest in functional packets of 4. Matter has 4 phases—plasma, gas, liquid, solid. Physics sees 4 forces—gravitation, electromagnetism, strong nuclear force, and weak nuclear force. The lower bubble also has its own packets of 4, and all of it is due to the fundamental clout of the Rule of 4.

The master code templated some lesser fractal variants that we find here in the upper bubble. One is the genetic code. DNA uses a polarized pair of pairs: the 4 base molecules. Thymine bonds with Adenine, and Cytosine bonds with Guanine. The molecules organize at a higher level into pair-bonded, polarized triplets. Their polarized molecular 6-packs are set in 64 (4 × 4 × 4) co-chaos patterns along the DNA double helix to generate fractal iterations of each species in a changing flux of specific details.

Another variant of the master code is the I Ching shorthand of polarized yin and yang. It uses a polarized pair of pairs: the 4 bigrams. Due to their polarized lines, two bigrams are stable, and two bigrams change. The lines organize at a higher level into pair-bonded, polarized triplets called hexagrams (polarized linear 6-packs) that describe 64 (4 × 4 × 4) dynamic patterns of events.

The I Ching's math shorthand relies on the numbers between 2 and 8, for that brief path of numbers can merge the linear, binary mode with the relational, analog mode to describe some analinear dynamic that charts a manifesting event by describing its basic fractal form, but not its specific details. The ancient Chinese described this dynamic patterning as the way of the Tao.

The West eventually came upon an independent discovery of the same math in the genetic code. Watson and Crick showed how DNA codes for the building blocks of organic matter. Its double helix holds the same system of 64 pairs of polarized triplets that appears in the I Ching hexagrams. Other so-called "primitive" societies—Egyptian, Babylonian, Toltec, Dogon—have also recorded other versions of these 64 dynamics. They are varietal riffs on the master code of co-chaos grown on a dp-tree.

Yet another variant is exhibited in your own psyche, which is far larger than your ego's territory. Your human mind is a tiny variant based on the universal mind. Your psyche's polarized pair of pairs is silhouetted in psychological type theory established by Carl Jung and then quantified by Katharine Cook Briggs and her daughter, Isabel Briggs Myers. So even your psyche operates according to elaborations of the polarized pair of pairs.

Every cell of your body carries a communication link with the universal mind, which is constantly updating each species. For instance, a human's invisible DNA record, if taken out of a single cell of the body and stretched, is over 6 feet long. The DNA record of life has been updating for 4 million years on earth, and for most of that time, life's task of deciphering those memos has been completely unconscious, arising naturally out of the rhythms on earth. More recently, science has been trying to decipher them in labs.

We humans also get personal memos sent to us from the larger unconscious in dreams—dream mail—that most people throw away unread since they cannot understand its forgotten language anymore. But it can be relearned.

Oddly enough, the people who are now taking our society toward the idea of a conscious universe are often the very same people who, since the Age of Enlightenment, have mostly been leading people away from that idea: scientists.

A new class of "psicientist" is rising in many branches of study that now seem to be converging in an uprush of discovery. We are realizing something that's both simple and profound: physics and metaphysics are converging on the idea that nature is alive and aware. If you suspect nature has some sort of a holistic intelligence, you'll find more than a few scientists in the same camp.

Rupert Sheldrake pointed out in *A New Science of Life* that we've had three basic attitudes toward nature. He summarized them in an interview with Jerry Snider in *Magical Blend Quarterly*, Summer, 1988: "Before the 17th century, practically everybody believed that the whole of nature was alive....a living god and a living world.... That was pretty much the universal view, not only in Europe but throughout the rest of the world. Then the mechanistic theory came along in the 17th century and said, 'Oh, no it's not like that at all. Nature is dead and basically inanimate, and governed by eternal laws made up by God who set the whole thing running in the first place....

"The vitalist's view grew up in reaction to this: 'All right, nature is dead as far as physics is concerned...chemistry and physics and geology and rocks and things like that are just inanimate, but living things are different. They are really alive....it is an intermediate position which accepts the mechanistic view of nature for everything except living organisms....'"

"The organismic view of nature—the one which I support myself and which many people are beginning to support in quite large numbers today—supposes that nature is alive; that nature is a living organism. In other words, it's a return, if you like, of the old animistic view of nature, but it's a new turn of the spiral at a new level of sophistication.... The organismic view says that nothing is just mechanistic, that the universe is an organismic universe that looks mechanistic only if you are looking at it through a mechanistic framework."

This TOE agrees. The universe is alive and well.

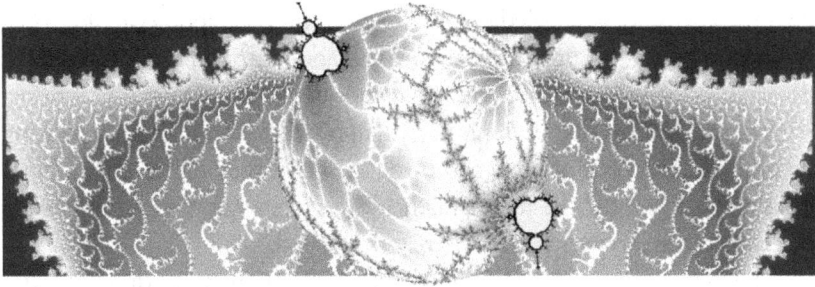

Chapter 6: Yin and Yang at Work

1. The I Ching in history

To understand the I Ching better, I went to live in China for a year. There I taught at Jinan University. Meanwhile, scholar Zhang Luanling tutored me weekly in the I Ching, helping me delve into its meaning and historical past.

I learned that Chinese tradition says the I Ching was first written down by King Wen (1152–1056 BCE). Its original text held the 64 hexagram yin-yang figures, plus their titles and hexagram judgments. The hexagram line meanings were added later by King Wen's fourth son, the Duke of Zhou.

The I Ching text had previously been an oral tradition. It was called the *I* (pronounced *Ye* or *E*). The title *I* had two meanings—*change* and *easy*, according to Fung Yu-lan in *A History of Chinese Philosophy*. The *change* aspect refers to the hexagrams' tracking dynamic change. The *easy* aspect refers to the fact that King Wen wrote the I Ching in ink on bamboo slats, an easier writing surface than the older, traditional surfaces of bones or turtle shells.

About 700 years later, the original I Ching text was expanded into a book by the Confucian-toned *Ten Wings* commentaries, thus expanding its title from just *I* into *I Ching*. (*Ching* means *book* or *classic*.) In the Tang dynasty, *Zhou* was added to the title to indicate that King Wen first wrote it down at the birth of the Zhou dynasty.

The text came into English as *I Ching* or *Book of Changes*. But most Chinese call it *Zhou Yi* (Pinyin transcription) or *Chou I* (Wade-Giles transcription).

As a document of the Zhou people, the I Ching refers to ancient names, places, and history. It uses analogies from a rural, agrarian, feudal society beset with tribulations under the ruling Shang dynasty. Some of those rulers are mentioned in the I Ching, especially Ti Yi, father of the last Shang ruler, Di Xin.

Di Xin was notorious for degenerate cruelty and extravagance that ended the Shang dynasty. It receded into history, visible only as brief mentions in the I Ching. Many Western historians supposed the Shang Dynasty was just legend.

In 1899, Wang Yirong, Chancellor of the Imperial Academy in Beijing, became ill with malaria. Medical advice sent him to an apothecary for a bag of "dragon bones" to cure the malaria. Wang and his friend Liu E realized the decayed bones were inscribed with what appeared to be pictographic writing.

Scholars learned that peasants in Henan province had long been plowing up oddly marked bits of bone and turtle shell that they called *dragon bones.* That name suggested the fragments held mystic energy without indicating exactly what that power was or where it originated.

The Henan peasants sold the bone and shell fragments to druggists who ground them up for medicinal purposes. Those "dragon bones" held markings whose pictographic communication evolved into Chinese script.

Archeologists heard about the inscribed bones and hoped for a major archeological find when they realized the peasants were tilling fields on top of the site of the ancient Shang capital. They began excavation and found Shang tombs, tools, and artful bronzes. The writing fragments contained at least 100,000 divination records made by the Shang dynasty (1600-1050 BCE).

According to an essay by Teja A. Jaensch, "It was discovered that the animal remains were primarily turtle plastrons and bovid scapulae dating back to the Shang era.…This city was primarily a ritual and sacrificial centre to maintain the wellbeing of the people and the land, where oracular rituals were carried out with the use of these shells and bones."

An ancient aura of power in the phrase *dragon bones* reminds me of the fractal shape below, which has been called a *dragon.* It is just one of many fractal shapes in the fringes of Julia sets around the Mandelbrot heart. This particular image replicates itself bilaterally in a polarized, flip-flop mirroring.

Fractal dragon

Scholars deciphered enough of the old inscriptions on turtle shells and bones to realize that their details corroborated a long-forgotten history of the ancient Shang dynasty, with many specific accounts of an era that had been dismissed and nearly forgotten except in legend...and in the I Ching text.

2. Symbolism of yin and yang in a Zhou mindset

Consider the mindset that is implicit in the ancient I Ching. Imagine what it meant to believe you live centered on the earth's foursquare landscape that is encircled by the round dome of sky. As you stand on the ground, you orient yourself by looking to the front, behind, left, right—establishing the four directions of North, South, East, and West.

Above you is the dome of heaven. It puts you smack at the center of things, no matter where you stand. This heavenly circle of yang energy expands to invisible abstraction, and meanwhile, the yin energy of foursquare earth holds you grounded in material reality. Snug in this land that will come to call itself the Middle Kingdom, you know you are obviously at the center of the world... or at least, of everything that matters.

This ancient mindset was eventually struck into the coin of the realm. The old Chinese coin below shows how in those days, even money sent a holistic, right-brained message. Its metal circle held a central square that symbolized the four corners of the earth covered by heaven's great dome. That way of thinking viewed yin and yang as complementary energies forever adjusting the balance of relationships within the Tao's ever-emergent flow.

Ancient Chinese coin

People around the world still use these old Chinese coins, not as money anymore, but to consult the I Ching. I prefer other methods...yarrow stalks, I Ching stones, or an accurate computer program based on the yarrow-stalk algorithm. However, I enjoy looking at the old coins for their shape and cultural significance. They suggest the mindset of a forgotten age.

To the ancient Chinese, the coin's circle symbolized the sky's lofty dome. By stretching so high, the sky illustrated the masculine principle of yang.

Imagine this with an ancient mindset. Stand on the earth and look up. Above you, the heavens rise over the foursquare expanse of solid earth stretching under your feet in all four directions. The dome's round, transparent volume is loftily inspiring, but the earth's steady, opaque foundation grounds your feet.

For yang energy, the Chinese chose a sky-high dome that invigorates us with its moving clouds, shifting winds, expansive reach. It holds the sun's moving fire that illuminates events by literally shining a light on things. Shafts of sunlight probe every cavity and crevice of the earth's folds to inseminate it with the motility of life. Its rays help us achieve goals and bring clarity to details.

To the ancient Chinese, that square hole punched in the coin's center symbolized the feminine principle of yin as the foursquare earth. Its steady, deep knowing plunges beyond the limits of bright yang logic. Only in the dark night can we begin to realize how yang's bright sunlight has blinded us to all the faint stars in their constellations beyond. Only when the ego's sharp focus relaxes enough to peer into the dark, shadowy hinterland of life's events can you view the analinear patterns of deeper meaning wheeling beyond the sun of linear logic.

A metaphor for yin energy is the soil itself. Earth has a stabilizing ability to contain things, due to its receptivity and acceptance of mystery. Earth symbolizes the mute truths that keep us grounded and oriented in life experiences. Earth is opaque, weighty with a dark capacity to hold seeds and give birth. Fertile, silent soil nourishes the seeds that sprout life from within its mysterious womb.

Equating earth with the feminine was a concept held not only by the Chinese. The English word *matter* comes from Latin *mater*, meaning *mother*, the *prima materia* that sustains us with its ability to open, receive, contain, birth, and stabilize. Holding steady beneath the bannering clouds and wheeling stars, the fecund earth bears us forth. Opaque and solid, it keeps us sensible, grounded, and "down to earth" with a gravitational pull so strong that our bodies cannot finally escape it even in death, for we are buried in the earth, turned into earth by cremation or time. Dust to dust.

Mother Earth births us, and then she buries us. Dark womb, silent tomb. Earth folds us back into itself...to be reborn anew? Thus yin energy can be calming, soothing, enriching. Or mysterious and scary. Traditional Chinese art often portrayed yin energy as a white, blue-eyed tiger. That fabulous, powerful beast is far more unaccountable, wayward, and unfathomable than the secretive house cat. Enigmatic and arcane, it symbolizes yin's mystery... like the earth itself, whose power is as vital as that of yang's dragon of heaven.

Thus to the ancient Chinese, earth symbolized the feminine yin principle; the sky above it symbolized the masculine yang principle. Yin and yang energy balanced each other, acting much like the two poles on a modern-day battery.

3. The binary sequence of trigrams

In ancient China, yin and yang math elaborated into a mandala that became a major cultural motif. The mandala's 8 trigrams or *ba gua* indicate the 8 compass directions. Its layout is reminiscent of the compass directions familiar to Western culture, but the 8 trigrams express a more refined polarity. Each trigram's rising stack of yin and yang layers show an increasingly complex polarization. This mandala still appears in Chinese shops and on household walls.

White Heaven
Father
S

Red Lake
3rd Son
SE

Green Wood-Wind
1st Daughter
SW

Orange Fire
2nd Son
E

Blue Water
2nd Daughter
W

Yellow Thunder
1st Son
NE

Purple Mountain
3rd Daughter
NW

Black Earth
Mother
N

Early Heaven mandala of trigrams in complementary binary sequence

Below is the Early Heaven order of trigrams sitting in a row. This trigram order puts the trigrams into a binary sequence from 0 through 7. In this series, we'll call it the binary order of trigrams.

0	1	2	3	4	5	6	7
Black Earth	Purple Mountain	Blue Water	Green Wood-Wind	Yellow Thunder	Orange Fire	Red Lake	*White Heaven*
Mother	3rd Daughter	2nd Daughter	1st Daughter	1st Son	2nd Son	3rd Son	Father

Early Heaven's binary order of trigrams

This trigram order sets the Mother, all six siblings arranged by birth and gender, and then Father into binary sequence along the row. Each trigram has its own separate identity and dynamic, from all-yin ☷ Mother Earth on the left to all-yang ☰ Father Heaven on the right. Their "bookend" placement suggests their parental power.

In the next chapter, which is more left-brained, we'll explore some of the mathematical significance in this binary trigram order. But in this chapter, I want to show you some social, poetic, and mystic aspects of the trigram symbolism that might otherwise get lost in the mathematical emphasis up ahead.

Each trigram's name, gender, and birth order sprang from an era when nature, agriculture, feudal, and familial relationships provided most analogies. An analog mindset likes to play with resonant possibilities, and ancient China loved to spin out a proliferation of poetic names in nature analogies.

But to a modern Western mindset, so many analogies can be confusing. It all begins to feel like too much metaphor, analogy, allusion veering into illusion. Nevertheless, careful exploration shows that each trigram's "personality" has its own distinct dynamic, some of which can be intuited by meditating on its name and traits. But some of it is quite subtle.

Always bear in mind that each trigram's chaos dynamic also carries an odd philosophic quality. Consider, for instance, the 1st Daughter: ☴ . The ancient Chinese called her Wood-Wind. A modern Western mind might fail to see any link between invisible wind and solid wood. But to the ancient Chinese mind, wind and wood share a remarkable trait.

First, consider wind. One cannot see the wind, only whatever it moves—waving grass, blowing dirt, rising smoke, flapping banners, driving rain, or drifting snow. The wind is invisible, yet it can carry tons of sand or snow in its unseen grasp; such power can etch rock bridges and build glaciers.

Likewise, the casual eye does not much notice how a sprouting tree grows larger from day to day. Cell by cell, its woody volume slowly swells from a tiny seed into a stem, a trunk, even into a tree as large as a banyan, cypress, cedar, or sequoia…all of it arising from wood's invisible, persistent strength.

To the ancient Chinese mind, both wood and wind had a certain dynamic of persistence that is symbolized by the unnoticed, inconspicuous oldest daughter, Wood-Wind ☴ . The 3 rising lines in ☴ describe her trait of persistence slowly becoming evident over time. This daughter's dynamic starts out with yin at the bottom of the trigram by appearing meek and inconspicuous. (Always read trigrams or hexagrams starting from the bottom line going upward.)

Yet over time, her succeeding yang lines declare that she is actually very strong, dedicated, and enduring. Wood-Wind tends to introvert and hide her power beneath a humble appearance. She is like a tiny acorn that gradually becomes an oak, or like the wind that over time sculpts a sandstone natural bridge. This shy yet competent oldest daughter creates lasting change through the gentle application of her self-effacing, persevering dynamic.

All of the trigram family members are powerful, each in a unique way. Each trigram's yin and yang lines depict its own chaos pattern, a specific dynamic symbolized by that family member's habits. If you pair-bond any two trigrams to make a hexagram, those two family members operate in a dynamic interaction that creates a larger, modified co-chaos pattern. In modern scientific terms, each trigram is a math symbol for the dynamic of a fractal chaos pattern. Bonding two trigrams into a hexagram produces the more complex fractal dynamic of a co-chaos pattern.

4. Working with the I Ching so it works for you

When you use the I Ching algorithm to ask a question, its answer offers you a hexagram whose yin-yang math shows a relevant co-chaos dynamic. That hexagram's mathematical figure and its accompanying verbal text describe a fractal co-chaos pattern relevant to your issue at hand. Its text will use ancient analogies from the experiences of a feudal, agrarian tribe in a long-ago culture.

Each analogy will tell you a story that you can endeavor to understand and apply to your own particular circumstances. Or not. It offers you holistic wisdom that is explained by analogies with a relational, qualitative emphasis rather than as data derived by quantitative analysis.

Since it is fractal, the hexagram's basic dynamic has a recognizable pattern, yet the specific details you're living out will be a unique variation on that pattern. In other words, your hexagram's basic dynamic is like a predictable container, but the load of contents you pour into it will vary uniquely in its specific details, according to how you are living it out. You can change those specific details according to how you deal with the pattern, and often you can even change one dynamic into another. You have that much free will to alter the course of your life.

Understanding the dynamic you are living makes you more aware of any options to live it better. You can alter your experience and often its outcome by your own free-will choices that fill in a dynamic's basic fractal form with the uniquely detailed contents of your own living iteration. It's your life.

To apply an answer from the I Ching algorithm to real-life events asks for the ability to consider it with both your left and right brain, using both logic and holism, seeking to find a balanced response that is complete and appropriate to the circumstances. It is this balanced approach that the modern mind often has trouble recognizing or honoring, much less doing.

If you are a natural philosopher, poet, or storyteller, it may come easily to you. If not, the I Ching may remain more or less opaque to you. Using it involves truing up your skill in pattern recognition, finding the right response to that pattern (fascinatingly, often by recognizing your blind spot on an issue), and then testing the success of your response in the emergent conditions of your own life's unique ongoing events.

I suggest that you determine whether or not the I Ching is something you can understand and apply in your own life. Maybe it will work out for you. If it does, sometimes you'll find (if you are like me) that it is not the oracle that is failing me, but I who am failing it. Occasionally I am not quite able to leap to the heights of its suggested rational response because of my own lingering sorrow, unresolved anger, insecurity, ignorance, or even my defiant, willful resistance to a reality that I've kept trying to ignore. Yikes! That can be regrettable.

But my efforts to understand a hexagram answer help me become more able to perceive what is blocking or aiding me. It really benefits me to dialogue with its dynamic flow, which I can by now recognize is actually seeing me and responding to my problem. Its algorithm is tuned to the master code itself.

Each of us has something unique to live out and offer. Too often, that gift gets lost in the muddles of misunderstanding, sorrow, pain, and regret. Learning through pain is hard. How much better it is to learn through the pleasure of cooperating with the watercourse way of the Tao. Then its meandering path that seems so frustrating to linear logic becomes time's inevitable gravitational flow toward truth. To tap into ongoing truth about yourself and the world gives a unique meaning to your own life and your reason for being.

Carl Jung remarked in his *Foreword* to the Wilhelm/Baynes translation of *The I Ching or Book of Changes* that sometimes a dialogue with the I Ching can clear the head and heart enough to open a path to more and better choices than otherwise would have been possible.

But remember that the I Ching has to be learned slowly. Using the I Ching is not about how smart you are, or clever, or canny. It doesn't let you enlist a battery of glib statistics and glittering bravura to parade in the bright lights of public hoopla, cueing that auto-applause so prevalent in today's modern culture.

It is more experiential than intellectual. Its hexagrams can gradually develop a webby fractal pattern of math and meaning for you. Studying it brings a quiet heart that is enriched by meditation and humility. Be patient with the trigrams and hexagrams, get to know them slowly, treat them with deep, intuitive connection as well as logical, linear thought, and then the I Ching may carry you into finding a meaning deeper than words can tell.

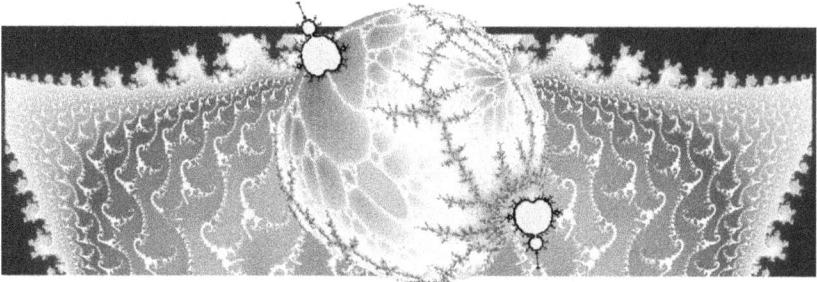

Chapter 7: The 2DD Triangle

1. How do you deconstruct a Mobius band?

Remember the 2DD triangle that we explored in Chapter 5? The 8-looping path around both sides of the 2DD triangle acts much like moving around a Mobius band or a Lorenz attractor's loops since the path keeps repolarizing as it switches from the space face to the time face…or vice-versa.

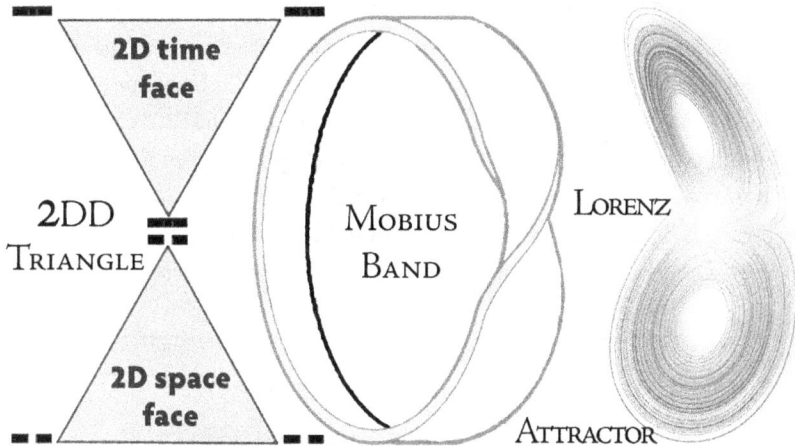

2D time face

2DD
TRIANGLE

2D space face

MOBIUS BAND

LORENZ

ATTRACTOR

All three dynamics use mirror-reversed polarity

As early as 200 CE, the Mobius band showed up on Roman mosaics, yet it was only in the 1980s that math began to cope with describing the Mobius band's shifting polarity. Mathematician Mikhail Gromov's work eventually helped others describe more easily how a pulse, a particle, or a planet moves in *curved* spacetime, where Newton's first law of motion does not apply.

Although Gromov founded the field that came to be called symplectic topology, it was physicist Hermann Weyl who named it by coining a word for its twisty mystery: *symplectic*—derived from Greek *symplektikos* meaning *braided together*. Weyl pointed out that a Mobius band has 3 traits: *motion, topology,* and *repolarization*. They braid together so smoothly that the eye cannot spot exactly where each trait begins, changes, or ends, so it becomes quite difficult to do the math on how a Mobius band repolarizes across dimensions as it twists.

Consider the Mobius bands shown below, adapted from "The Hidden Twist to Making a Möbius Strip" by Kevin Hartnett in *Quanta Magazine*.

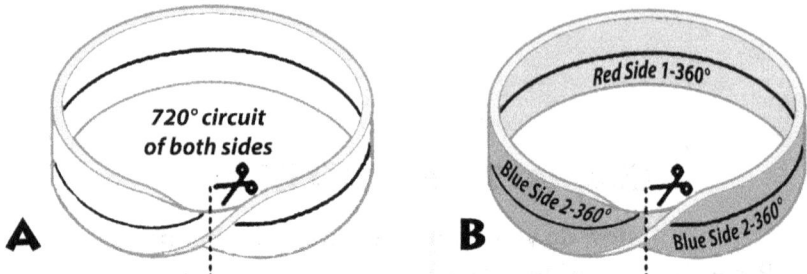

A-Two 360° circles on the white band—**B**-Snip 2-color band at its transition point

A-Make a Mobius band out of white paper. Draw a straight line along both sides of all 720° on the band. Where does the line's first 360° orbit close and its next 360° orbit open? You can't tell since the band keeps twisting as you turn it.

B-Make a new Mobius band from a strip that is "polarized" by having a different color on either side. Now it's easy to tell where the 360° orbit of one color stops and the 360° orbit of the other color begins along the total 720° circuit.

C-Now snip through your band at that transition spot between the two colors. Your cut removes the strip's twist, so it can lie flat on a table. A different color polarizes each side of the strip, and both sides still show that straight line you drew. But where did its mobic twist go...dimensionally, I mean?

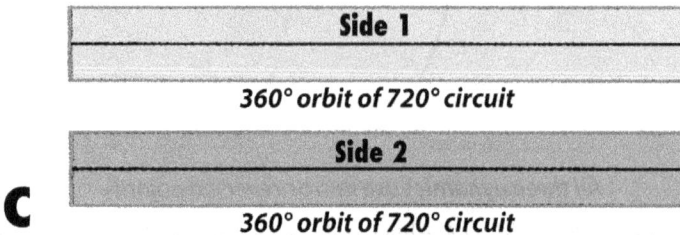

C-Your flat strip has 2 sides with 2 lines

Symplectic topology argues over how to identify, explain, or measure this twisty mobic event. In "A Fight to Fix Geometry's Foundations" in *Quanta Magazine*, Kevin Hartnett said, "In flat, Euclidean space...motion can be described in a straightforward way by Newton's equations of motion. No further wrangling is required. [But] In curved space like a sphere, a torus or the space-time we actually inhabit, the situation is more mathematically complicated."

In the 1980s, mathematician Vladimir Arnold formalized symplectic math by declaring that whatever moves along a vector field and then comes back to the same angular position where it started makes a *closed orbit* with a *fixed point*.

In the 1990s, mathematicians Kenji Fukaya and Kaoru Ono sought to refine how to count those closed orbits and fixed points. But there were problems, and mathematicians Dusa McDuff and Katrin Wehrheim ignited a years-long debate by challenging the foundational rigor in symplectic geometry.

McDuff noted a ground rule in *What Is Symplectic Geometry?*: "Symplectic geometry is an even dimensional geometry. It lives on even dimensional spaces, and measures the sizes of 2-dimensional objects rather than the 1-dimensional lengths and angles that are familiar from Euclidean and Riemannian geometry."

That ground rule of symplectic geometry applies at the mobic scale. The 2DD triangle at the mobic scale exists in even-dimensional space *and* in even-dimensional time, plus it also has a repolarizing twist in the tension path running around it. Count the pulses going around the 2D time face and the 2D space face. Do you count 6 or 8 pulses? As Dusa Mcduff's paper noted about symplectic geometry, "...one of its most intriguing aspects is its curious mixture of rigidity (structure) and flabbiness (lack of structure)."

2. A new/old math is needed here

Symplectic geometry counts closed orbits and fixed points. But the Luo Shu and Magic Square count the 4 polarized pulses that open/close each orbit around either face of the 2DD triangle's cdp hull ⋉. The **5**-spot's polarizing warp at the center marks the 5th/10th/15th/20th *whatever* transition... however long you keep orbiting the Luo Shu, Magic Square, or 2DD triangle.

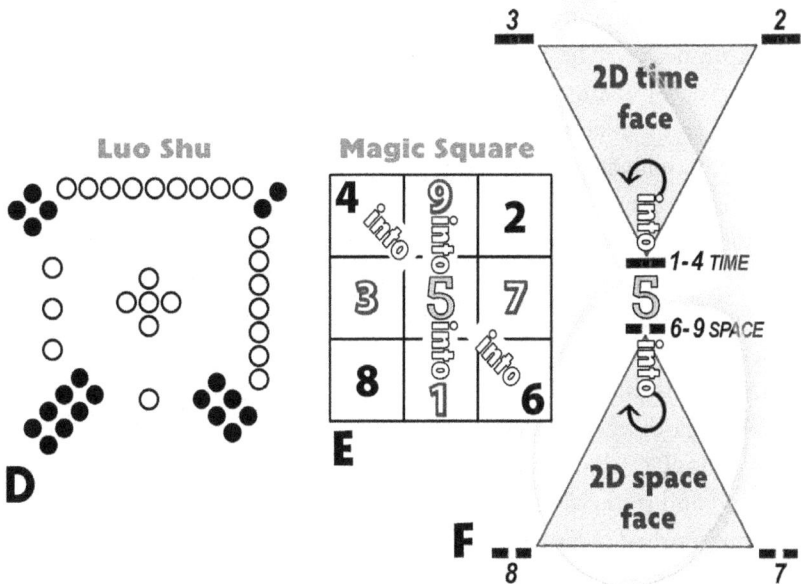

D, E, F-*The Luo Shu, Magic Square & 8-looping the 2DD triangle*

This 8-looping, polarized path on the 2DD triangle has some aspects of a Mobius band and some aspects of a Lorenz attractor. Thus this TOE proposes to call it the *mactor* dynamic that can only happen at this scale of the universe.

So how many poles did you pass as you orbited around both faces? Trace the 8-loop again if you're not sure. Find the counter-clockwise symbol ↺ and start your finger following around the time face at poles **1**, **2**, **3**, and **4**. Then the central **5**-spot closes the time-face orbit. It also opens the space-face orbit where the clockwise symbol ↻ takes the path around poles **6**, **7**, **8**, and **9**. Again, the **5**-spot closes that orbit. You can orbit both circles of the 8-loop repeatedly by jumping through the central **5**-spot each time you reach it. This repolarizing passageway of the **5**-spot between the two domains of 2D time and 2D time closes an old orbit and opens a new orbit as long you keep traversing the 8-loop.

If you want, you can trace out the same number path that you just trekked on the cdp hull ⧓ also running your finger around the Luo Shu (Image **D**) or the Magic Square (Image **E**).

Note, you are counting poles, not corners. With each 8-loop, the numbers run as **1**, **2**, **3**, and **4**…then the central **5**-spot offers a passageway to **6**, **7**, **8**, and **9**…and the **5**-spot again becomes the pass-through point. Orbiting on this path, you'll keep iterating the same sequence of numbers again and again, jumping through the **5**-spot's passageway each time.

This is how the Luo Shu counts out the pulsing pathway around both faces of the 2DD triangle, and at the center, it always adds a 5th beat, 10th beat, 15th beat…. The central **5**-spot offers a warping passageway between two domains. It marks the repolarizing site on the 2DD triangle.

Trekking the 8-loop path around the cdp hull ⧓, the two farthermost pulses on either face iterate a same polarity. They propose the two stable bigrams: the time face proposes ▬▬ , and the space face proposes ▬ ▬ . But right above and below the repolarizing **5**-spot sit two poles with different polarities. Those two poles propose the two changing bigrams—either changing yin ▬ ▬ or changing yang ▬ ▬—depending upon how you approach them while trekking around the 8-loop.

Image **G** shows how the sequencing of polarities around both faces of the 2DD triangle develops the 4 bigrams: ▬ ▬ , ▬ ▬ , ▬▬ , and ▬ ▬. They are the I Ching's (and dimensional latticing's) foundational polarized pair of pairs.

Build and read each changing bigram while trekking along the 8-loop. You must move down past the 4-yin pole on the 2D time face into closure's **5**-spot *before* you reach the 6-yang pole on the 2D space face. Poles **4** and **6** make both changing bigrams. Since bigrams are built and read from the bottom up, moving downward through the **5**-spot sequences the two poles as ▆▆ ▆▆. Moving upward through it sequences the two poles as ▆▆▆. This trek around both 2D faces proposes all 4 bigrams, a polarized pair of pairs: ▆▆ ▆▆ , ▆▆ ▆▆ , ▆▆▆ , and ▆▆▆ .

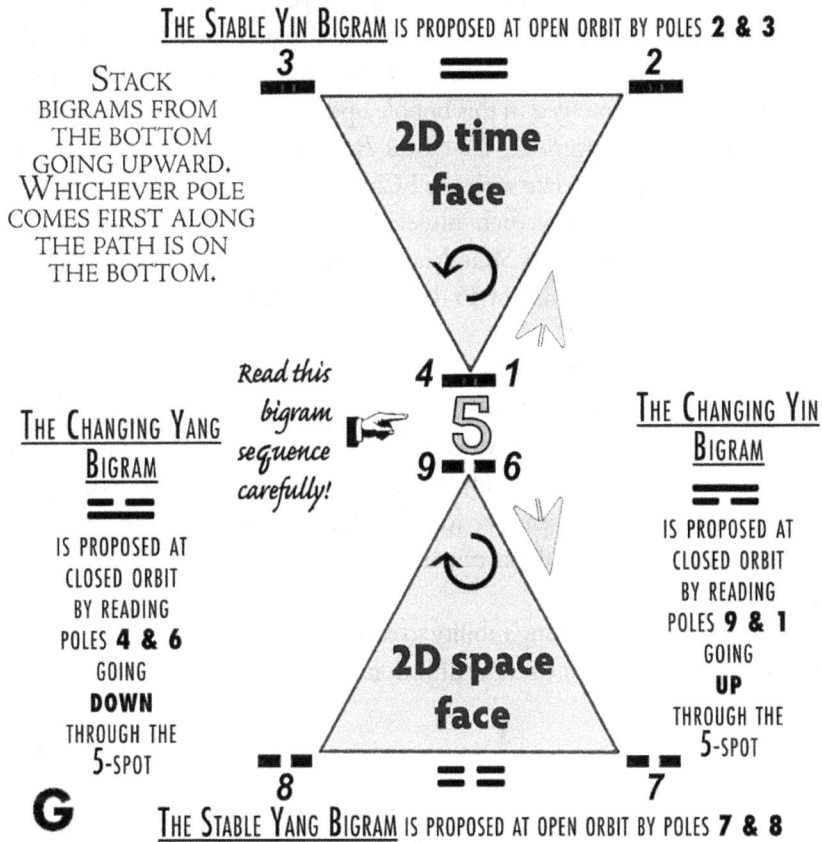

THE STABLE YIN BIGRAM IS PROPOSED AT OPEN ORBIT BY POLES **2 & 3**

STACK BIGRAMS FROM THE BOTTOM GOING UPWARD. WHICHEVER POLE COMES FIRST ALONG THE PATH IS ON THE BOTTOM.

3   ▆▆▆   **2**

2D time face

Read this bigram sequence carefully!

4 ▆▆▆ **1**

⑤

9 ▆▆▆ **6**

THE CHANGING YANG BIGRAM

▆ ▆

IS PROPOSED AT CLOSED ORBIT BY READING POLES **4 & 6** GOING **DOWN** THROUGH THE **5**-SPOT

THE CHANGING YIN BIGRAM

▆ ▆

IS PROPOSED AT CLOSED ORBIT BY READING POLES **9 & 1** GOING **UP** THROUGH THE **5**-SPOT

2D space face

G

8 ▆▆ ▆   ▆ ▆   **7**

THE STABLE YANG BIGRAM IS PROPOSED AT OPEN ORBIT BY POLES **7 & 8**

G-*The 2DD triangle's 8-looping path proposes the 4 bigrams*

The traffic sign summarizes the 4 bigrams. They sit on the traffic sign's 4 corners. The hinge between the 2 triangles is the 5-spot's repolarizing dynamic.

Does the orbiting path of tension around a 2DD triangle ever really stop? No. As it closes the orbit on one triangle's face, warping polarity sends it on a new journey orbiting around the other face. This polarized flip-flopping between both faces will continue to the end of time and space because that's what makes them.

You may be wondering why I am bothering to show you similarities between the ancient Luo Shu's Magic Square and the 8-looping double orbit around both sides of the 2DD triangle. It is not a random oddity that they mesh so naturally. Recall that this series' underlying premise is the idea that a foundational master code established certain mathematical patterns in fractal co-chaos that recur again and again in variations across every scale of this universe.

If you've read the previous three volumes, you already know this TOE says the genetic code and I Ching math are two mathematical variants of the same underlying paradigm, a foundational master code of information from which gravitation emerged in this bubble upon reaching the quantum scale.

For instance, Volume 3, *Co-Chaos Patterns,* showed how DNA's 64 molecular 6-packs correlate with the I Ching's binary order of 64 hexagram 6-packs. It showed how each molecular 6-pack of DNA can add up mathematically to the Luo Shu's Magic Square of 15. It showed how the ancient Chinese He Tu map of 55 dots correlates with the 55 atoms in the 4 base molecules of DNA

These and many other parallels were pointed out to establish the premise that we have, in effect, a Rosetta Stone with two known codes and one unknown code. The two known codes are the genetic code and the I Ching math that can shorthand it. Studying their parallel traits can help us decipher the obscure, sketchily-seen, yet foundational master code that templated them. Cross-correlating them can help unlock the universe's foundational master code.

In Image **H**, the I Ching's ability to shorthand complex layers of polarity led the Chinese to use it to show directions on ancient maps.

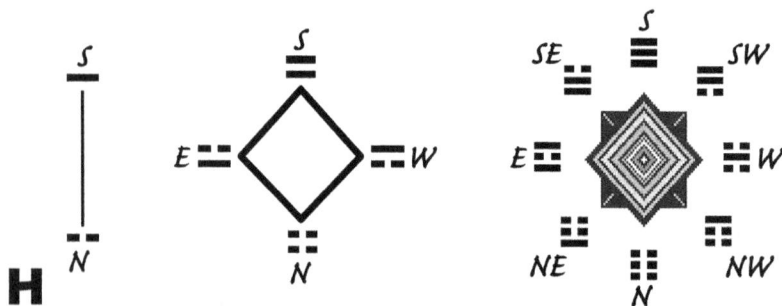

H-The 3 stages of polarized development show exponential growth

In Image **H**, Chinese maps traditionally put South at the top, North at the bottom, West on the right, and East on the left...the exact opposite of a modern Western map. A basic map on the left shows North as yin - - and South as yang ▬ . Reading rightward to the central image of a diamond, more layers of polarity develop as the 4 bigrams indicate the 4 directions of North, South, East, and West. On the far right, a mandala uses 8 trigrams to indicate

8 compass points that include NE, NW, SE, and SW. These 8 trigrams sit in Shao Yong's binary order of Early Heaven with reversing mirror-image pairs.

By the way, did you notice that the compass points increased exponentially as $2 \times 2 \times 2$, or 2^3? All 8 directions tap into the 2-to-8 number sequence that is so essential to co-chaos patterning, as you saw in previous volumes.

Because bigrams are of special interest to us here, Image **i** below shows how the 2D diamond organizes its bigram polarity. The exploded diamond on the right reveals that each diamond segment is tipped by a yin pole and a yang pole.

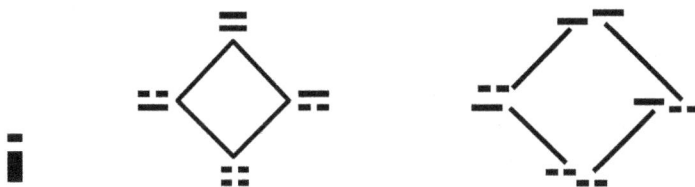

I-*Ancient Chinese compass polarized by the 4 bigrams*

A simple algorithm sorts out each diamond segment's polarity, reshuffling them into the 4 bigrams. You can sequence all 8 poles of all 4 bigrams by going clockwise around the exploded diamond and examining each exploded diamond segment in turn. On each segment, its lead pole sits on the bottom of the next bigram, but its rear pole sits at the top of the previous bigram.

3. The path around the cdp hull jump-starts our arrow of time

The tension path around both faces of the 2DD triangle at the mobic scale is the polarized precursor to what will eventually generate our own upper bubble's arrow of time, and likewise, the other bubble's arrow of space. However, while its dynamic is still just an ultra-tiny 8-loop orbiting a 2DD triangle at the mobic scale, its route merely traces a polarized, looping path in 2D space, then in 2D time, then in 2D space...you get the idea.

But when the cosmegg develops enough dimensional surety to generate many hourglass cells ⧗ projecting 3D space above and 3D time below, they merge holographically into mirror-twin bubbles conjoined at the mobic-scale interface. Each hourglass cell holds 3 polarized, 8-looping paths that merge into the tensor network of a single, polarized dimension 8-looping across both bubbles.

In the upper bubble, our tiny, diverse, electronic minds know the upper half of all those 8-loops as the omnipresent arrow of time moving in 3D space, pinioning us in endless *now*. However, the lower bubble's giant tachyonic mind moving in 3D time is unified by its omnipresent arrow of space fostering endless *here*. It is hard for us to imagine or understand the particulars of how that other bubble's 3D time and space arrow operate since it defies our own bubble's experiential reality by a mirror-inversion of our own bubble's properties.

As each 8-looping path rotates in each hourglass cell across both bubbles, it is in effect describing 2 orbits in a dynamic that was born at the mobic scale on the ultra-tiny 8-loop of a single 2DD triangle. Its archetype still persists in gravitation. For instance, when women run, their breasts move on an ∞-loop.

We will continue to explore how 2DD events at the mobic scale developed the projection of two complementary bubbles of 3D space and 3D time, plus the ubiquitous tensor network constantly circulating across both bubbles, with both orbits in its 8-loops polarized as the moving arrow of either time or space.

I first saw this dynamic of mobic-scale dimensionality in a life-changing dream on March 4, 1985, and for several years, I had no way to understand or describe it. Then synchronicity tweaked a German friend to give me Martin Gardner's book, *The Ambidextrous Universe*, and I had an open weekend to read it while I was attending the Jung Institute in Switzerland.

There I came upon Gardner's advice to make a Mobius band and play with its polarizing property. So one snowy afternoon in 1987, I sat at my desk in Zurich playing with Mobius bands while something kept ringing in my head like a crystal chime. I heard *Ting! Ting! Ting!* But not with my ears.

Something was coaxing me to consider, "Hmm, this mobic band is really just a 2D plane caught on an endless warping path. It's like a weird, endlessly twisting, rotating Flatland." Then, tired of cutting paper strips to make bands, I grabbed a printer sheet of pinhole-feed paper and tore off a margin strip. On both sides of that strip, I penciled in crosses dangling from its punched holes. Then I twisted the strip and taped it into a Mobius band.

I was fascinated to see that by rotating the band, I could polarize what it carried. During the first 360° of rotation, each punched hole and penciled cross formed the ancient Greek symbol for Venus: ♀. But during the second 360°, Venus flipped upside down to form the ancient Greek symbol for Earth: ♂.

By rotating my warped circle of paper, I was constantly repolarizing the information it carried. I could switch the ancient symbols for two planets back and forth by repolarizing them to point up or down. Or to look at it another way, I was turning the negative – side into the positive + side. Or yin — — into yang ——. Or *no* into *yes*. Or north into south. Or space into time.

4. The 2DD triangle gives foresight & foreplace

Playing with the Mobius bands helped me clarify what that dream showed me back in 1985 about space and time dimensions. The Luo Shu even offered me a way to conceptualize the problem of 6-8 number indeterminacy.

I first stumbled upon the 6-8 conundrum in that 1985 dream sparking this series, as told in Volume 1, where I contemplated us seated at the table as numbers in a shifting dynamic: *In my bird's-eye view above the table, I watch this 6–8–6–8 pulsation going back and forth. (Only 10 years later did I realize that the 6–8–6–8 dynamic foreshadowed my work on how bigrams propose and confirm the mobic-scale co-simplex geometry in the co-chaos paradigm.)*

That dream's starkly dramatic dimensionality now somehow reminds me of a panorama I saw back in 1970 when my husband and I were visiting an amusement park atop Mount Tibidabo, tallest mountain in that range along the Mediterranean coast, and overlooking Barcelona. John disliked carnival rides, so I was riding alone on a Ferris wheel atop Mount Tibidabo. As I sat in the sunlight on the Ferris wheel in my swinging little car, I watched a dark thunderstorm raging in from the ocean and starting to stretch over the city.

Ferris wheel in a storm

Lightning was striking repeatedly down over the city. It was thrilling to watch, yet also scary to realize that the storm was moving this way. Hey, lightning might even strike here before I could get down off the Ferris wheel!

But in this weather, I reminded myself, surely the park manager and employees knew how to judge any danger to their Ferris wheel riders...yes?

As I sat in the sunlight, watching lightning play over the city, I could also hear grand music blasting from loudspeakers fixed to the outer walls of a stone church maybe 45 meters/50 yards away. It was the *Temple del Sagrat Cor* (Temple of the Sacred Heart). That huge, dignified old church was broadcasting at full volume the words and music of its Easter Sunday mass to the crowd milling around outside on the amusement park grounds below me.

My Ferris wheel car seemed to vibrate with the rolling thunder, with the many-voiced chorus, with the monumental organ booming out over the crowd. I thought, "What a dramatic view of contrasts! It's worthy of El Greco! What an amazing country. You can attend church on a Ferris wheel!"

Here is a 2012 Google satellite view of the top of Temple del Sagrat Cor. If you look closely in the lower right corner, you see the shadow cast by the Ferris wheel that I rode on an Easter Sunday.

The top of Temple del Sagrat Cor & the Ferris wheel on Mt. Tibidabo

The church itself looks like a Romanesque fortress made of tan stone that is weirdly topped by a second church, a neo-gothic cathedral made of gray stone.

Temple del Sagrat Cor viewed from the Watchtower inside the park

5. 2D time & 2D space give foresight & foreplace

Now back to a deeper music in a lower register…the polarized pulses that sketch in the 2DD triangle. Its polarized tension runs around 2D space, back to home base, then out again to encounter the Opposite Other, 2D time.

This 2DD triangle generated at the mobic scale employs a math whose effectiveness can be resolved down to four bigrams. For us, counting out the Luo Shu's Magic Square consolidates the 2DD triangle's opened/closed orbits, polarity warps, and pulse-counting into the four bigrams, organized as a polarized pair of pairs, that drum the music of creation along its 8-loop.

Playing with paper Mobius bands eventually helped me understand my dream. And I think the ancient Luo Shu counts out the 2DD triangle's polarity shifts at the mobic scale better than symplectic geometry could. It helped me count out the tension path orbiting around one side of the 2DD triangle, going 1-2-3-4 in a constant closure of the old orbit...*warp-5* to the other side...while also embracing the new orbit on that endlessly renewing 8-loop path. The polarity events in those orbits can be shorthanded by the four bigrams, a polarized pair of pairs integral to the life of the universe itself.

Can more dimensionality fit onto this sturdy triangle at the mobic scale? Indeed yes, because 2D space and 2D time act as a dedicated team giving the cosmegg foresight and...well, foreplace. That tiny 2DD triangle exists literally a micro-breath beyond the singular points of *here* in space and *now* in time. The cosmegg's elegant ability to bestride 2D space and 2D time at the mobic scale gives it a real advantage. It can foresee and foreplace how to grow more elaborate dimensional structures yet remain fail-safe. Stuck here in the upper bubble without having 1D time, much less 2D time, we cannot imitate or even imagine that ability very well.

As a consequence for the cosmegg, any further construction has a much better chance of determining which way to grow safely, efficiently, effectively— space-wise and time-wise. The encoded trait of foresight and foreplace in 2D space and 2D time will allow the cosmegg to apply the Rule of 4 in a new way that will develop an even more complexly polarized yet failsafe dimensionality, thus escaping the binary trap of *either/or* that becomes a dead end.

The binary, linear mode insists on a forced choice of *either/or:* "Go big or go home." But the analog mode amends that forced choice into the *both/and* option of "Go big *and* go home. Repeatedly." By merging both binary and analog modes into the co-chaos paradigm, the cosmegg can develop an analinear math that brings optimal results with economy of effort, plus giving it enough leeway to allow error and yet maintain integrity in the Double Bubble body.

Hmm, if this triangular truss is strong enough, accommodating enough, then construction could be adapted on the fly to support whatever feasible step turns up next. Why not just push the limits and see what happens? Rise on up to the heights of possibility and then take the plunge that commits to reality!

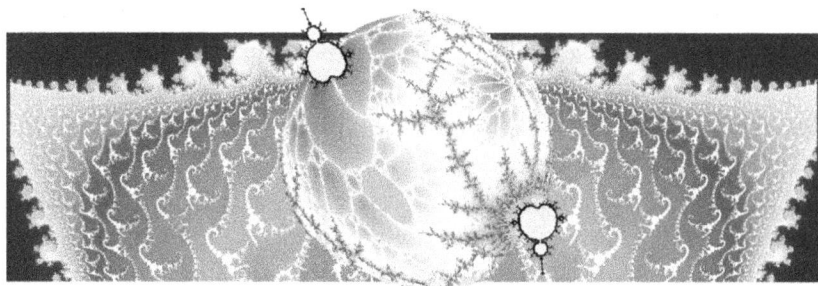

Chapter 8: Finding the Third Way

For at least two thousand years, a philosophical preference divided our globe culturally. The West emphasized an *either-or* tendency. The East emphasized a *both-and* tendency. It philosophically split the cultures of Earth, rather like a huge global brain whose Eastern hemisphere honored a right-brain approach, while its Western hemisphere honored a left-brain approach.

Each hemisphere touted its own *modus operandi* and misunderstood, misjudged, and even disparaged the other hemisphere's cultures. But both hemispheres are now opening many connective pathways to integrate the world's population into a global culture that shares the best in human ways of thinking and living. We have a lot of issues to juggle. Languages, politics, finances, religions, races, genders—all can become touchpoints for misunderstanding, misjudgment, disparagement, even war. Can we rise to a new, higher order in our individual minds, in our culture, in our globe? Can we embrace this Earth in a new culture of diversified unity among peers?

1. Chinese writing as pictorial analogies

A fact often unnoticed but quite profound is that speaking different languages actually cultivates different mindsets. The Chinese language, for instance, is quite attuned to analogs. Its written characters rely on implied relationships so much that until the 20th century, its text even had no punctuation; in effect, it was all just one sentence. People had to subdivide the characters into sentences by intuiting the context of parts within the whole.

The Chinese language still has no pluralization of nouns in the English sense, no verb conjugation, no articles, nor even prepositions or conjunctions, unlike most Western languages. Instead, a web of meaning, highly nuanced and pregnant with connotation, bonds each word to its neighbors in context.

The grammar of negation, for instance, is much more keyed to a particular query, and its response will mirror that specific question being asked. Thus each sentence in Chinese quivers with analog resonance and associations.

Analogy is embedded in the brushstrokes of written Chinese. Its script originated by symbolizing objects. The first principle of character writing was "Imitating the Form." For instance, this is *man*: 人. See his silhouette? He has lowered arms, and his legs are striding across the ground. That character imitates his form.

To portray more abstract notions, a second principle developed called "Pointing at the Thing" or "Indicating the Thing." As an example, here is the man again, but now his arms are outstretched to symbolize the character *big*: 大. He is showing us the measure of something big…perhaps he is describing a fish that got away. You see how a slight twist of the brush could modify an image to suggest a new twist in meaning.

Another example is *mouth* 口. It is a square opening drawn with the loose flow of a brush rather than the sharp nib of a pen. You can put dashes above the mouth to show a sketchy version of an extended tongue that is wagging with sounds as it talks. Those dashes modify *mouth* into the more abstract idea of *word* 言.

Turning the pictogram for *mouth* into *word* demonstrates how China evolved a written language by modifying what was originally a simple imitation of form using pictorial analogies. How does modification indicate something so abstract as *trustworthiness* or *good faith*? Why, let's have that man stand by his word, literally, to honor it. Here is *good faith*: 信. It is formed by the man moved into a radical position to support his word and stand fast by it.

That last example illustrates a third principle of Chinese writing. It may be translated as "Joining the Meanings." It links one picture with another picture to suggest a new condition. Here, *man* joins *word*. Western languages often do something similar to create a striking image—for example, English says, "He's leapfrogging the issue"—but our Western metaphors put their images into words built by alphabet, not in pictures inked by brushstrokes.

Although Chinese writing began with the imitation of form using pictorial analogies, it eventually developed six basic principles that include a phonetic harmonizing of sounds and a borrowing of sounds between some characters. Thinking in Chinese promotes a non-Western way of processing language. Laboratory testing showed that the Chinese pictograph system activates different parts of the brain from those energized by alphabet-based Western languages.

Chinese characters cannot signal their pronunciation as alphabet-based words do. However, as a plus, people from opposite ends of China can read the same written characters, even if they speak that shared language in dialects so widely differing that they sound almost unintelligible to one another.

Tonality is another analog characteristic of Chinese, and indeed, of most languages in Asia. Tonality uses various pitches of voice to differentiate among its words. Mandarin Chinese uses four pitched tones plus a "toneless" tone.

In Chinese, you must say a word with just the right tone of voice to indicate the right meaning. For instance, only the tone of your voice will reveal which meaning you intend for the syllable *ma*—meaning *mother, hemp, horse,* or *scold*. If you use the wrong tone for *ma*, you indicate the wrong word...and, unfortunately, tone was often my *bete noire* in speaking that language.

Why did tonality develop? Because it had such utility. Ancient Chinese was made up of monosyllables, which imposed a very limited range of sounds to designate meanings. Even today, Mandarin Chinese has only about 400 possible syllables, while English has about 12,000. Thus in Chinese, varying the spoken tones of those 400 syllables can help to differentiate various meanings.

And since so many words in Chinese sound very similar or even alike, people who are conversing will frequently sketch out a character's image with a finger on the palm so that the listener can literally "get the picture." That doesn't happen with alphabet-based languages.

A vocabulary with many words of like or similar pronunciation also brings far more frequent punning. The Chinese do not regard puns in the same derisive way that Westerners do. An analog brain dotes on puns, while a linear brain often groans. Chinese, in fact, plays sound against image with a poetic dexterity that fosters a heightened sensibility to imagery in general.

A student in my graduate class in China once told me that he wanted to name his child "Listening to the Sound of Rain in the Pavilion." He said his fascination with the name lay not just in its condensed symbolism in Chinese but also in the mellifluous unpacking of its evocative imagery, where someone is being sheltered from the rain, yet listening to that soothing white-noise in nature.

"Listening to the Sound of Rain in the Pavilion" is an incredibly long personal name in English, but not so in Chinese, what with dropping out all the prepositions, articles, and even that unnecessary word *sound*, since it is automatically implied if you are listening to rain. Most of the name's meaning is implied, not found in its actual characters.

Traditionally, analog societies have tended to name their children, cuisine, art, homes, and even streets with an imagery that is rooted in nature's rhythms. But for Westerners, nature imagery in children's names is no longer in fashion currently. In fact, names that invoke nature's images are currently regarded as quaintly unfashionable—for instance, Pearl, Ruby, Iris, Forrest, Garland, Running Deer, Daisy, Dale, Eagle Feather, Spring, Ellwood, Violet....

Western family names usually refer not to nature, but instead to an occupation: Cook, Carpenter, Smith, Taylor, Brewer, Baker, Turner, Draper,

Dyer, Hunter, Mason, Parker, Tanner, Fletcher, Sawyer, Carter, Marshall, Skinner, Fisher, Miller, Farmer, Porter, Potter, Archer, Clark (meaning clerk), etc.

2. Foods & festivals reveal the mindset

Chinese culture still embraces the analog in myriad ways. Food, for example, is eaten with a pair of chopsticks that are brought into continuous relationship—using the analog *both-and* style of pincer action. Traditionally, the Chinese preferred to apply those chopsticks while gathered in a group at a table (preferably round), enjoying the bond of food together. The group shared central dishes in common as they chose a bit from this platter of poetically named *Dragon and Tiger* (snake and cat) or that bowl of *Phoenix Feet* (chicken claws).

Even nowadays, especially at feasts, Chinese food is traditionally precut, so you do not bring the struggle and disharmony of divisive knife action to the table. Everyone reaching into the shared serving bowls for tidbits creates a continual repositioning in the food dance. The chopsticks weave a ballet of flux that acknowledges everyone's place in the warmth of the sharing group.

People usually discuss the food in detail—its appearance, flavor, preparation, purchasing, and so on. People joke about food, remember past meals, other moments connected with food. The table flows into a *participation mystique* as the culture's children gather around the great analog mother for dinner.

Westerners, on the other hand, make different, individual requests for food in a restaurant, and they eat off unshared plates in a statement of implicit territoriality. Each person consumes a personalized portion of food that is set down in its private bailiwick of plate and placemat. Each person employs the weapon-like knife and fork, using a cut-and-thrust act of *either-or* division rather than the *both-and* pincers of chopsticks. Westerners also argue, debate, and criticize more at the table, perhaps because they are less committed to the group identity and rapport-building quality of sharing communion in food.

Currently, the basic identity unit in China is mostly the group, unlike the individual unit that is preferred in the West. Analog connectivity to group identity in China outweighs the discrete personal identity, so individual freedom in China often becomes less important than group unity and security.

Chinese folk events offer a feast of data for the cultural anthropologist who is exploring its analog, relational mores. Custom still employs the lunar calendar to designate those major holidays closest to its collective heart—the Spring Festival and Mid-Autumn Festival, both named for seasons of nature. Both are three-day celebrations tied to the lunar calendar.

In the lunar calendar, the moon is like a goddess of the night; she receives and reflects the sun's light in a silvery glow. To ancient China, the moon symbolized yin energy based on the waxing and waning moon, thus suggestive of female cycles. In fact, in most cultures, menstruation is identified with moon cycles. Although the lunar calendar is still used in China for its major traditional holidays, the solar calendar adopted from the West is applied for ordinary work weeks, fraught with the precise scheduling of business's data-choked details.

In contrast to China, most major American holidays are based on significant individuals (always men)—for instance, Christmas, Easter, St. Valentine's Day, President's Day, Columbus Day, MLK Day. But some of those honorific holidays were repurposed from much older, more analog, nature-based celebrations such as the winter solstice (Christmas) or spring equinox (Easter).

3. Eastern style...both-and!

In ancient China, yin and yang were partners, not opponents. Both poles needed each other to exist in a state of *both-and*, not *either-or*. This attitude was evident in Chinese thought, so that the mutually exclusive, shunting gates of binary thinking did not gain dominance over analog networking.

Ancient China was not enamored of the sharply defined ego stance that began to assert so much personal identity in the West. Even now, many Chinese still hold onto the Marxist-Stalinist view that sees psychology as a Western pseudoscience. Many still do not separate themselves much from the rhythms of nature or the *participation mystique* of the group. They do not notice the borders of the oceanic unconscious because they still live so much inside it.

The Chinese philosopher Zhuangzi [Pinyin] (369 BCE - 286 BCE) expressed such a mindset in this anecdote: "Once upon a time, I, Zhuangzi, dreamed that I was a butterfly, fluttering here and there, a real butterfly, enjoying itself fully without knowing that it was Zhuangzi. Suddenly I awoke, and came to myself, the real Zhuangzi. Now I do not know whether I dreamed that I was a butterfly, or whether I am now a butterfly dreaming that I am a man."

By not seeing life as a binary war between yes and no, right and wrong, pain and pleasure, good and evil, China managed to avoid the Western cultural divide between lofty mind and sinful matter. Joseph Needham put it this way:

"Europeans suffered from the schizophrenia of the soul, oscillating forever unhappily between the heavenly host on the one side and the atoms and the void on the other; while the Chinese, wise before their time, worked out an organic theory of the universe which included nature and man, church and state, and all things past, present, and to come."

For a long time, the Chinese mindset brought a pacifying balance to its culture, along with a slow, steady progress. The country was seen as the Middle Kingdom at the center of the world. From border to border in that huge land, its people were encouraged to look down on foreigners as beings less subtle, less cultured, less evolved and aware. They viewed "foreign devils" as unpleasant ghosts, due to their white skin, eerie blue eyes, and diseased-looking yellow hair.

In fact, Owen Lattimore suggests in *Studies in Asian Frontier History* that the Great Wall of China was built at least as much to keep Chinese culture contained and coherent in its way of life, as it was to wall out nomadic barbarians. Walls both physical and psychological shut out the perturbing dissonance of stray cultures. But in that very superiority lay the seeds of decay.

Inside the Great Wall, China created its centrally sanctioned inventions and very slow, steady rate of development. China became Confucian, honoring a stability of social form over the older Taoist tolerance for quirky latitude that erupts into creativity. Occasional invaders from the North and West were absorbed into the Han Chinese populace. Eventually, China became so set in recycling its habits that it forgot how to permit change. It could no longer innovate without threatening its very identity, the nation's way of conceiving of itself. Feudal China became so ultra-stable that it sank into inertia.

By 1650, Chinese technology had lost out to the driving competition of the West. As the West dropped feudalism and connected capitalism to scientific and technological developments, its progress jumped well ahead as its people eagerly turned abstract theories into scientific experiments in many areas.

In *The Evolution of Chinese Science and Technology*, Jin, Fan, Fan, and Liu (all four are graduates from universities in China) took a rueful view from their perspective. Regarding 2,500 years of technical achievements in Chinese and European history, they describe Chinese science as showing plenty of early promise, but then becoming outstripped by Western progress in the long run.

Why? I would say it was because Imperial China stayed awash in the cyclic, analog tides, turning symbolism into superstition, becoming imprisoned by a great wall of isolation. It codified the masculine and feminine roles into the collective sexist ideals of war lord and foot-bound lady. Eventually the government itself took on the yang persona of dominant war lord, while maintaining an iron-handed rulership over the restricted, compliant yin masses.

Up to 1800, the submissive yin populace remained isolated from the rest of the world, like some enormous foot-bound lady kept hobbled and subdued at home by her yang lord and master. China's people remained feudally dependent upon the rigid government for their definitions of individual, group, and national identity. A person could not challenge the severe laws without receiving unsparing punishment. John Fairbank says in *The Great Chinese Revolution: 1800 to 1985*: "Authority figures should not and could not be challenged. Criticism endangered authority and was therefore unacceptable."

My opinion is that the ritualized yang/yin roles forgot to honor the essential dots of the opposite in the tai chi symbol (☯. As a modern-day metaphor for wholeness, it suggests that we recognize, honor, and develop the contrasexual complement within. We all carry both yin and yang energy. The skill is to develop both, and then use each appropriately as needed in events.

The analog mindset that honored attunement to the whole was an ideal of harmonious accord, but over time, it distorted into kowtowing to the yang emperor by a yielding yin populace, submissive to their place in the social order, until over 4,000 years of dynastic China, it slowly stultified progress.

4. Western style…either/or!

Meanwhile, on the other side of the globe, Western culture did the opposite of harmonizing into inertia. Its many diverse countries kept it so far from equilibrium, in fact, and so dynamic in those tensions that fluctuations amplified into structure-breaking waves that accelerated the frequent escapes into higher orders of organization on the logical side—but with chaotic feedback erupting from the repressed networks of relational, analog qualities.

Western cultural development crested on the rise of logic and male superiority in Greece, and then on the triumphant armies of the Romans. But after the fall of Rome, there was no longer a common government, writing, or road system such as existed continuously over thousands of years in China. Thus, Western civilization did not mode-lock into the analog bonds at any cost that characterized a unified, dynastic Chinese civilization, which rose in a continuous gentle curve of development.

Instead, European culture went more binary, emphasizing the *either-or* mindset that championed competition and conquest over the subtler nuances of analog relationships. Former Roman territories split, then split again, with minor kings and feudal lords driving the cycles of boom and bust.

The heroic goal became a will to win, succeed, come out on top. Triumph over nature. Trump others. Conquer, subdue, or destroy women, dusky races, the physically or mentally impaired, witches, infidels. Deny one's own sin, flaw, whatever smacked of a hidden, dark, analog domain in one's own unconscious. In other words, combative yang energy won out over yin. Consciously, that is.

But as yin energy retreated into disapproved or piously repressed layers of Western culture, its energy no longer appeared just dark and mysterious. It warped into dangerous, sinful, forbidden, misshapen, wrong, bad. Public disapproval forced many to act out their repressed impulses in artistic or sexual expression, for such kinds of creativity faunch at being suppressed.

During the Crusades, men sought in foreign lands the Grail of a silver chalice—such a yin symbol!—even as their women were locked away in chastity belts back home. A huge priestly population forswore the sinful daughters of Eve—at least publicly, that is—yet they fathered many illegitimate children. Alchemists sought to transmute lead into gold via the inspiration of a shadowy, usually publicly unacknowledged *soror mystica* or mystic sister.

Europe's many small nations were hostile enough to develop different languages and cultures, yet still close enough for frequent cross-fertilization. This fostered continually differentiating factions that kept competing and jousting for individualistic clout among the small fiefdoms and principalities.

The Renaissance era, especially in its arts, brought a creative merger of yang and yin energies, and the West surged well ahead of China in science. Frequent fluctuations among the many and diverse European nations created competitive rifts that drove exploration, theories, discoveries, inventions. This tension-packed system sparked new learning centers, technologies, and laws that struggled to keep up with so much change. Alvin Toffler's book *Future Shock* chronicles that accelerating rush of technological innovation.

Exploration also birthed the discovery of new continents, and soon the allure of a huge green virgin called the New World lured the really yang types to adventure. They either decided to leave home or they had it imposed on them. Here came the Pilgrims and Puritans, armed with Bibles and righteousness. Here came the second sons, rebels, indentured servants, criminals, outliers of all sorts from the population of those more settled and content to stay at home.

By the 1600s, the competitive heritage bred into the Western collective psyche developed vying cultures that exhibited many characteristic *either-or* traits. For instance, the economic style of the West is acceleration and slump, swinging wildly in its flip-flops of yang boom and yin bust. The populace myopically focuses on just one-half of that polarity at a time—either boom or bust. For instance, in the global financial meltdown of 2008, Geoffrey Wheatcroft noted in the *New York Times* that, "Bertie Ahern, the Irish prime minister at the time of the rapid economic growth, merely boasted, 'The boom is getting boomier,' preferring to unknow the truth that booms always go bust."

"Boom-and-bustiness" became a catch phrase in economics, yet it was originated by physicist Robert May. That phrase suggests a "booming" male expansion that subsides into the troublesome "bustiness" of a dark specter of bad times in vaguely female form. Historically, the flip-flops of boom-and-bust describe the economies of Austria, England, France, Germany, Greece, Holland, Ireland, Italy, Poland, Portugal, Spain, and the Americas.

In 2008, that instability convulsed the West enough to affect the global market economy, even entraining in its boom-and-bust swing a formerly immune Eastern hemisphere. Its perturber was the enormous tension generated by that deep, greedy, prey-and-predator split in the West's collective mindset, where its combative poles of *either-or* choose competition over cooperation.

5. Finding a new balance for East and West

In the West, those who have so much often want even more, and never become satisfied. This "control over nature" has culminated in a technological threat to our environment, to our species, to other species, to the planet itself.

But finally we in the West are beginning to realize that acting so resolutely *either-or*, linear, goal-oriented, bottom-line, is dangerous business. We are starting to realize that when we conquer nature, we turn her into a raddled whore left crippled by the culture that rapes and sells her body. To have your way with nature, to subdue it, control it, possess it, own it, destroy it—that is a Western mentality on steroids. A foolish goal. In having it, we get had.

The Eastern hemisphere, meanwhile, becomes continually more aware of the economic benefits in promoting personal identity, cultural exchange, and freedom to speak up. After living in China, I admire so much about the Chinese people. I hope for their opportunity to embrace the best of their past in their future, while also opening up to the best of the West. Essential to that development is overcoming a fear of losing the safe containment symbolized by an autocratic ruler, a fear that is still evident in present-day North Korea.

The East is reaching out to the West for its technical skills, its higher standard of living, its mod-pop zest. But the East should also reckon the margin between Western benefits and its bad habits. Unfortunately, thousands of years of healthy, eco-friendly ways are falling against the brutal blows of de-naturing technology. To me, this is a major pitfall that the East must seek to avoid.

Meanwhile, the West reaches eastward. It wants more analog connection, more sense of pulling together as one instead of maintaining an arbitrary, adversarial posture of relentless debate, conflict, rivalry, one-upmanship. To achieve that, it must quit raping nature brutally. It must learn regard for nature, even human nature, and hold it in embrace rather than take it in plunder.

Techno-weary Westerners these days may grow starry-eyed over the mysticism of ancient China or the tea ceremony of feudal Japan. Both convey a sense of attunement to cycling rhythms, of living nestled in the *participation mystique* of connection with nature, even with the eternal verities…a sense of embracing reality, integrating all, digesting all, surviving all.

Sounds grand, doesn't it? But yin acquiescence into the flow has its own dangers. Absorbing the increasingly ritualized procedures without much question or protest, overwhelmed by that weight of accrued cycles, they shoulder history's burden of the ancestors, government, traditional rules for everything. Everything! For ancient China, for feudal Japan, so much yin, analog containment eventually became too cloying an embrace.

A young Chinese man once told me bitterly as he painted (mild) graffiti on a wall, "The trouble with us Chinese is that we'll accommodate to anything." He realized that his culture's analog embrace did not necessarily make it a kinder, gentler society. Just a more unified, tractable one.

Make no mistake, the unmitigated analog embrace can be brutal to its people, its culture. It can breed despots, repression, saving face, and the unquestioned custom of its people becoming a living sacrifice to the gods of ritual. For millennia, analog cultures sacrificed the people's welfare to custom.

Even at the start of the 20th century, an Ashanti tribe in Africa sacrificed 100 women and a few men when their god-king died. His whole harem was singled out for the honor. Willing victims gathered at a holy banquet and got drunk. Then female executioners came in and strangled with leather straps the many widows who assumed that as the honored dead, they would accompany the god-king into the beyond, where their happiness would bless the tribe.

Analog cultures get swept up in the *participation mystique* that fosters a hive mind buzzing with glib mottoes and gossip-fed attitudes. Its collective identity has overwhelming tides of favor or disfavor. Grudges, bribes, gangs, sanctions, and feuds. Lucky numbers and auspicious word-plays. Spells, omens, and good luck charms. Vengeful ghosts, demons, and spirits.

To me, this is the upcoming danger that the West must face as, first, it dips a toe into the seductive swell of analog vibes, then jumps naively into oceans of emotion scaled toward blanket judgments and mob mentality. It has not yet found its sea legs, much less a sea-worthy boat. I hope for a coming ship of state that balances individual rights with group good, that honors both linear smarts and analog wisdom, that cares about how we live on this living Earth.

This series addresses some intentional ways to combine linear logic with analog insight at the personal level. Learning to do that, society-wide, could bring about the transcendent third condition of stability plus evolutionary thrust that is modeled in the co-chaos paradigm itself. After all, it works in our genetic code, and in our Double Bubble universe. They both are designed for the evolutionary long haul. I like good things that last a long time.

How about you?

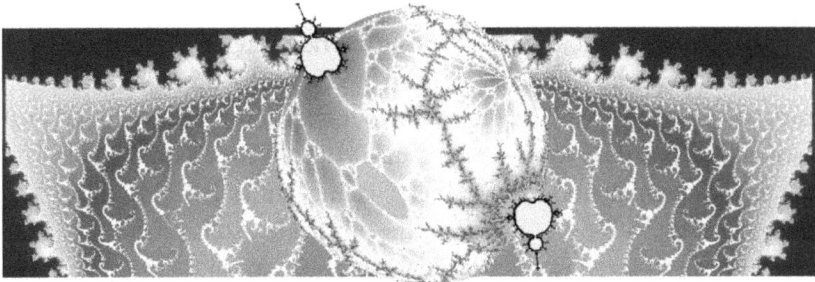

Chapter 9: Tetrahedral Triumph

1. Taking an overview of dimensional growth

Up here above the quantum scale, science today says our space is 3D, time is an arrow, and never shall the twain become equal partners. But far below the quantum scale, way down in the mobic scale where space and time emerge, they can and do and must become equal partners to develop a 2DD triangle.

Recall from Chapters 5 and 7 that at the mobic scale, the cosmegg's pulsing alternations of nonbeing and being—symbolized here as 0-1—used an *on*-pulse to establish a 0DD point. Then two *on*-pulses established the tension of a 1DD line polarized by space and time—symbolized here as -1 and + 1. Then three *on*-pulses established the tension of a 2DD triangle with two faces. One face is polarized by 2D space, and the other face is polarized by 2D time.

Those steps have differentiated and refined the cosmegg's polarity of space and time in 3 rising orders of dimensional complexity. The graphic below sums it up using I Ching shorthand since it is succinct, easy, and effective. You see it on our space-time shapes of co-simplex geometry below.

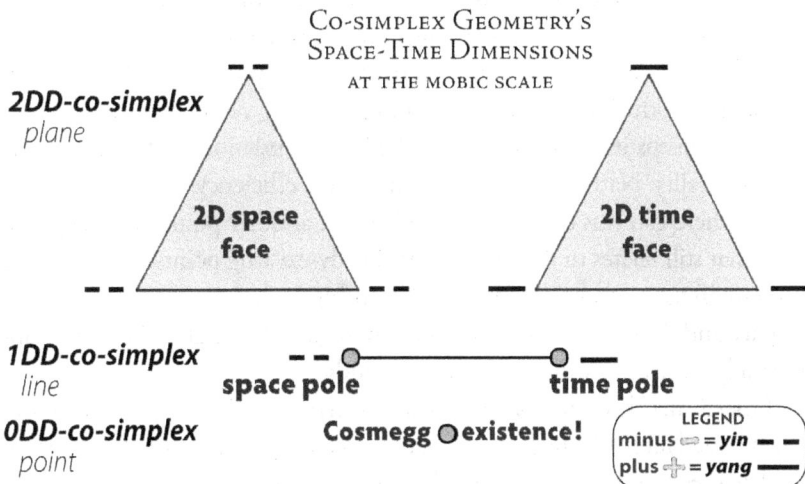

CO-SIMPLEX GEOMETRY'S
SPACE-TIME DIMENSIONS
AT THE MOBIC SCALE

2DD-co-simplex *plane*

2D space face

2D time face

1DD-co-simplex *line*   **space pole**   **time pole**

0DD-co-simplex *point*   **Cosmegg ○ existence!**

LEGEND
minus ⇨ = *yin* — —
plus ⊹ = *yang* ——

Three rising stages of dimensional polarity

The cosmegg's ongoing pulses of being and nonbeing are gravitational beats that can maintain the 2DD triangle faced by 2D space and 2D time. A tension path is ∞-looping—or 8-looping (take your directional pick)—around each face, and its polarized tension can be maintained. Tallying up the pulses on the tension path around the cdp hull $\bigtimes$ counts out as **1-2-3-4**...*warp*-**5** to the other side...**6-7-8-9**...*warp*-**5** to the other side....on and on.

Now the cosmegg wonders, "How can dimensionality grow next? Why not leap somewhere really new? Why not apply the Rule of 4 in yet another way?" The cosmegg is en route to establishing the universe's two bubbles, and to watch that happen, let's take another ride on the dimensional roller coaster. Let your inner eye act as a camera snapping videos through the dimensional trip, recording images to enter in the cosmegg's baby book.

After a fourth beat closes the **6-7-8-9** orbit around one triangular face, the **5**-spot *warp* does not land back at the old **1-4** base on the other triangular face. Instead, it seeks a new spot to land yet remain within the mobic-scale limits. But if it stays on the same plane as the 2DD triangle, adding another point would turn it into a 2DD square, parallelogram, trapezoid, or some other form of quadrilateral...and that just won't work.

"Why not?" you may wonder. "Isn't there still enough room in the mobic scale? Let those other three points scoot closer and rearrange themselves to make a plane with four corners. Square the triangle, so to speak."

No, at this scale, they cannot get any closer and still be different. And no, a square or other parallelogram cannot suffice here, anyway, for a triangle is already the simplest, strongest, and most versatile shape for a 2DD plane. There's no gain in creating another 2DD plane that merely has more corners.

Moreover, the existing 2DD triangle already offers the double whammy of two polarized faces with a sweet mactor dynamic that proposes the four bigrams reconciling both sides. So the developing cosmegg vetoes any quadrilateral shape and keeps its 2DD triangle as the best foundation for building more dimensionality. Nature resolves it down to that: efficiency.

So where can this questing fourth beat set another point in a really new place that still relates to the three other already-existing points? It must relate to them; in fact, it must maintain a relationship with them that is equidistant in space and time. Yet it cannot do that on the 2DD triangle's plane. Its shape is already complete, and its points won't budge.

Question: Oh, cosmegg, why not just turn back? Why not be satisfied with a 2DD-sized life? Why must this dangerous leap of growth absolutely happen?

Answer: Analinear drive was bred into the cosmegg when its original nothing became something that alternated nonbeing and being in *off-on pulses* of linear, binary 0-1 to make a point, as meanwhile, the 0 of nothing also split into that analog, relational, *yin-yang* polarity of -1 and +1 to make a line. The two modes merged into real clout: it is linear and binary, yet also analog and relational, and it drives for more dimensionality in its relentless search for growth.

That fourth beat is so motivated that it cannot squelch ambition and resign itself to running back home on the same old circuit of **1-2-3-4**...*warp*-**5** *to the other side*.... No! It cannot stay stuck in that hypnotic, endless double-orbit!

On the other hand, it cannot leap totally away from the mobic scale and abandon that 2DD triangle already settled there, for all pulses of information are family born of the same ilk, generated as polarized variations that develop dimensionality within the mobic-scale strictures.

So the new location of a fourth pulse must inevitably exist in polarized kinship with the other three, no matter how much it is a renegade upstart, a pipsqueak, the odd man out insisting that the other beats diverge from their smug, triangular, 2DD stability and open up to a wider latitude that includes it.

How can the Rule of 4 help the cosmegg grow bigger? Become more than just a micro-verse without much scope, guttering out like all those already-failed, wannabe universes dying around it?

Analinear drive presses the cosmegg insatiably onward to find new options. After all, it has already parlayed nothing into a 2DD triangle, and the potential in 0 is still nagging the cosmegg to sort out more possibilities. Space, time, matter, energy—all are jostling to manifest at higher levels of dimensional order.

But *what* order? That fourth beat must land, not back at the old **1-4** home base, but instead, find a new plane of spacing and timing but still within the confines of the mobic-scale limits.

2. A tetrahedral solution?

Survival instinct kicks in. The ornery fourth beat resolves not to stick to the old rut and go back home. With encoded 2DD foresight and foreplace, it sees when and where to attempt something really drastic, a new fourth point.

The beat twists away from the plane of that old triangle, and still at the mobic scale, sets its fourth beat in a new location of spacing and timing... and wow! Its new point achieves an equilibrated relationship with at least two points in the previous triangle to form a new 2DD triangle at a different angle!

No, wait! That new fourth point is generating not just one but four triangles! They're flashing quickly in and out of existence. Again and again! This new point multiplies a single 2DD triangle into four 2DD triangles, each with two faces. Their points bond in the most economical, equidistant relationship possible as they flash by at the mobic scale, due to 2DD's encoded gif of foresight and foreplace.

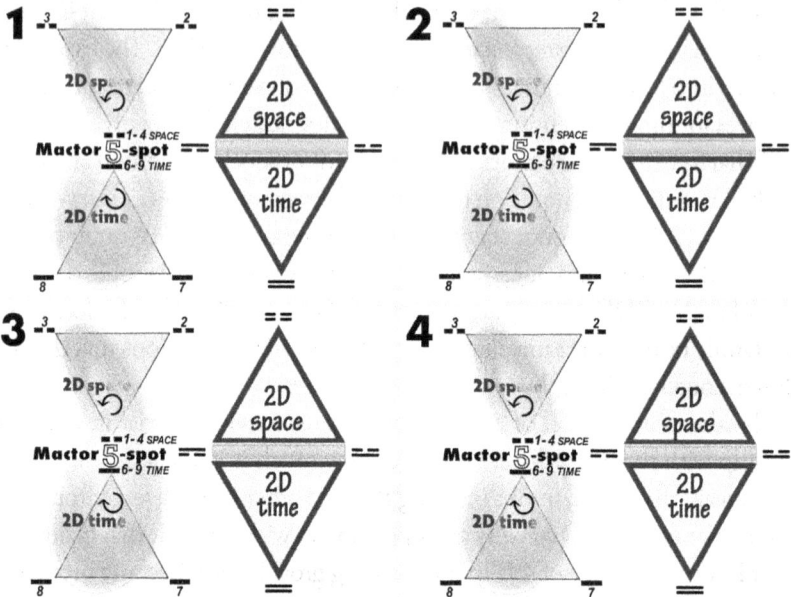

1 | **2**
2D space / 2D time — Mactor ⑤-spot 1-4 SPACE / 6-9 TIME — 2D sp, 2D time

3 | **4**

4 polarized 2DD triangles come from just 4 points

Here are all 4 2DD triangles splayed out to show both faces beside their traffic signs bigrams. You recall, the mactor path around a single 2DD cdp hull ⋎ proposed the 4 bigrams: ☰ ☰ , ☰ ☰ , ☰ ☰ , and ☰ ☰ . Bonded together at their corners into a geometric shape, all 4 of the 2D triangles form....

3. A tetrahedron confirms the bigrams that the triangle proposed

...Aha, a tetrahedron! Regard this tetrahedron made of 4 2DD triangles.

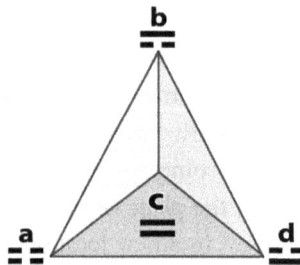

The 4 bigrams bond together the 8 planar surfaces

Each triangle's 2 faces of 4 dimensions work in sync. The tension running around each triangle's two faces recalls the 2 polarized orbits of a Mobius band or a Lorenz attractor. It's the *mactor* dynamic, so important that it energizes the tiny scale where space and time emerge, and so powerful that it eventually will throw the arrows of time and space across the Double Bubble universe.

The 4 polarized 2DD triangles form a tetrahedral shape, and their 2DD planes bond at all 4 corners, whereupon their complex polarity consolidates into a polarized pair of pairs: ⚏ , ⚎ , ⚌ , and ⚍ . The 4 inner surfaces and 4 outer surfaces meet to confirm their bigram bonds.

These bigram bonds ensure the tetrahedron's structural reliability as all the 2DD triangles merge their complex polarities into just 4 bigrams sitting at the tetrahedral corners. Bigrams symbolize the complexity of information in + and − polarities that bond together the 4 polarized 2DD planes whose 8 inner and outer surfaces meet to form a tetrahedron. It's a new twist on the Rule of 4.

And look, growing this new dimensionality develops so many different riffs on the 2-4-8 progression that's available in 2^3! All of this meets the Rule of 4 admirably, providing a new, higher-order dimensionality that still maintains the analinear number cooperation enabling our universe to exist.

4. A rumble over the property of volume

So much for the tetrahedron's 2DD *planes*. But what about its tetrahedral *volume*? How does the volume act? Space and time must somehow divide the tetrahedron's volume in an equitable way that maintains the Rule of 4. But how?

The cosmegg inspects the tetrahedral pyramid shaped by 4 different 2DD triangles, looking for the best way to take advantage of this amazing architecture that offers the intriguing new trait of volume. But where and how to express that volume? And what exactly would this remarkable tetrahedron consider its volume to be? Would it be 3D space? Or 3D time? Which primal gets to claim the grand prize of wonderful expansion into greater dimensionality?

Both space and time strive to claim the volume jutting into a new dimension. They're brawling over who gets to control the volume. This is dimensional war! Who will win? Will it be space? Or time? And if one primal claims the volume, that leaves the other primal reduced to…what?

Maybe you suppose, "Oh, the solution is obvious! It's clear to science that the universe developed dimensionality into 3D space and an arrow of time. That's why we now have this expansive, gracious 3D space that our pop-out, molecular bodies are tinkered to inhabit, but time got reduced down to the painful arrow of an enemy we can never quite conquer."

Yes, judging by what our science assumes, the winner must have been space. From here, it looks like space won 3D volume, while the loser was reduced to a thin arrow that current science calls 1D time, although its arrow cannot go both ways, but instead shoots along on a one-way trip flying constantly ahead through space's voluminous 3D gut. Talk about shafting your partner!

But why doesn't such unbalanced dimensionality make the universe short-lived? Wouldn't it fail from a lack of that essential symmetry so often demanded by good physics? Surely a universe so lopsided would exist only for an instant and then wink out due to the asymmetry of its spacetime signature that stands ungainly and unsteady, like a wounded bird balancing on just one foot.

Oh, wait! According to current science, that unbalanced posture already *is* the standard signature for our experiential spacetime. Science does not yet recognize that space and time begin down at the mobic scale, not up at the quantum scale. And it certainly does not yet notice that this is a stone soup universe, enhanced by the cosmegg annexing some add-on universes back when they were all at the formative stage. (See Volume 6, *Stone Soup Universe*).

Anyway, the infinitesimal cosmegg at this juncture is still just the size of one tetrahedron made by repeated iterations of its 4 adjoining 2DD triangles, each of them faced by 2D space and 2D time. That makes the cosmegg very small indeed. No atom, electron, or quark could fit into it.

The mobic scale can easily hold all the flat triangles quickly flashing by at whatever angle is required to bond each using 2DD's foresight and foreplace. But the mobic scale cannot hold the result of all that bonding: *volume!* This scale cannot contain volume, yet the tetrahedral shape insists upon it! Volume cannot fit into the flashing, quick-shifting 2DD reality of planar triangles. In those dimensional limits, volume has no way to express itself. What to do?

Don't give up hope now, cosmegg! One step at a time. Hindsight says there is hope, for we humans do exist, and so much inchoate potential still struggles to rush out and find fulfillment in a universe that works.

In the mobic scale, space and time are stuck in a quarreling quandary. Each primal wants the 3D volume. Likewise, neither primal wants to be relegated to the role of a mere arrow acting as solitary, shafting villain!

The cosmegg struggles to resolve this *either/or* conflict. It juggles options, trying to find a miracle fit for volume. Never mind about developing mattergy yet, nor those other forces that are awaiting beyond gravitation.

The cosmegg surveys the tense situation in its ultra-tiny innards and issues an edict: okay, space and time, you've got to divvy up this new property of volume equitably. Whatever happens, it must occur in an even-handed fashion that keeps a steady balance going within the whole system. It must happen symmetrically, and it must somehow obey the Rule of 4.

But how to do that? The leap to make tetrahedral surfaces was easy enough. That act could establish all 4 of the tetrahedron's 2DD triangles with their 8 quick-flashing, planar surfaces, inner and outer. But the tetrahedral *volume* is still up for grabs. Volume is inherently 3D, so an equitable division into 1½ D space volume and 1½ D time volume would be impossible and also futile, for it would lose the remarkable attributes of contiguous 3D volume.

Yet the cosmegg also realizes that in a truly balanced universe, neither space nor time would commandeer the 3D volume for itself alone, leaving the other partner bereft, stuck with just one dimension of either time or space... although it would honor the Rule of 4 (that is, if the 3D space:1D time ratio claimed by current science actually did hold true).

The cosmegg faces two horns of a dilemma, and choosing either horn will gore the Rule of 4: *Horn 1*—subdividing the 3D volume vitiates its utility. *Horn 2*—if either space or time wins the 3D volume, an appalling fact remains, rendering that victory moot. Neither result is good enough! Neither option can provide the symmetry so essential for actually, truly balancing out the universal stance of dimensionality so wobbly on one foot.

Does this unresolved contest between space and time for the 3D volume indicate that the cosmegg must drop back to the ignominy of a single 2DD plane? Give up exploring this amazing gift of volume offered by a tetrahedron? Will its impasse result in a quick retreat back to life as a 2DD mini-verse without hope for growing its potential? Or even a stymied death for the cosmegg? Probably, if it had stayed stuck in a binary, *either-or* mindset. But the math paradigm underlying the cosmegg is analinear, able to go big *and* go home.

View the tetrahedron's 4 points trying to cope with the issue of tetrahedral volume like 2 couples golfing their way around a course. Or like a card table where a foursome sits partnered as two competitive duos intent on winning a bridge tournament. Or like North, South, East, and West working together on a compass. Then you have an idea of the interactive dependence embedded in this new, tetrahedral shape dealing with the Rule of 4 and the number commitments that are involved in each polarized pair of pairs.

5. Can the universe maintain the Rule of 4?

Carl Jung said the major dilemmas of life are fundamentally insoluble if you remain stuck on the level where they originated. You must rise to a higher level of order that gives enough perspective on the points at issue to see how to resolve them. The cosmegg now must rise to a higher order of dimensionality that somehow honors the polarized pair of pairs in the Rule of 4, yet also resolves the dilemma of space volume vs. time volume.

You've already seen the cosmegg leap to higher orders of polarized organization several times before. You saw it emerge from sheer nothing by realizing its *nothing* as the number 0, then pulsing in and out as 0 and 1, and also polarizing that 0 into the -1 and +1 tension of a 1DD line sketched by space and time poles.

We can use a p-tree (polarized bifurcation tree) to show dimensionality's first fork of *minus-plus* polarity that makes a 1DD line. We symbolize that polarity in I Ching shorthand on the p-tree as *yin* — — and *yang* ——— .

Next, the cosmegg rises to a higher order of dimensional complexity as the p-tree develops a second level of polarized forking. Its more complex polarity of bigrams was first proposed by the tension path around a 2DD triangle, and it is now confirmed on the points of a tetrahedron made of 4 polarized 2DD triangles.

2ND FORK :   4 BIGRAMS

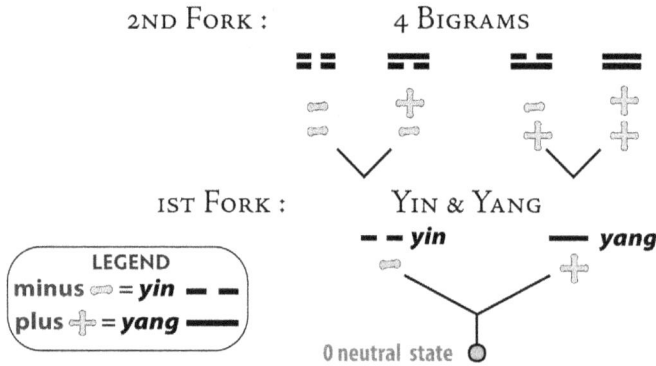

1ST FORK :   YIN & YANG

LEGEND
minus ⊂⊃ = *yin* — —
plus ✛ = *yang* ———

0 neutral state ◯

4 polarized paths = the 4 bigrams at the 2nd Fork of the p-tree

The p-tree's 4 branches hold a polarized pair of pairs. Its binary symbols count, reading left to right, from 0 through 3: ☷, ☳, ☶, and ☰ . The bigrams may also be arranged to sit in several other orders. Here are the stable pair ☷ and ☰ ...and the unstable pair ☳ and ☶ . Here are the yin-based pair ☷ and ☳ ...and the yang-based pair ☰ and ☶ .

The 4 bigrams are compact, flexible, versatile, enigmatic, and they honor the Rule of 4. (We normally read I Ching math figures from the bottom up.)

6. Clout of the tetrahedral archetype

The cosmegg still has not resolved its tetrahedral volume quandary. Hindsight says it got resolved and so well that we can even spot the tetrahedral shape's enduring clout, for instance, in the carbon atom, the basis of all life on Earth. Why is carbon the basis of organic life here? It's because the carbon atom often constructs tetrahedral-shaped molecules shape to gain versatility, and it always uses tetra-friendly bonds that allow it to go to interesting extremes.

Here's an example of carbon's extremes: diamond and graphite are both made of pure carbon. Although they are chemically identical, their crystal structures are very different. Diamond is hard, strong, shiny, clear, and abrasive. It insulates electricity. Graphite is soft, weak, dull, gray, and lubricating. It conducts electricity. Such difference is due to their varying crystal structures.

Diamond is super-strong because each carbon atom spokes out to 4 other carbon atoms at a 109.5° angle. If you set lines of tension connecting them, you get a perfect tetrahedral shape whose molecules pack tightly together. This gives diamond a rigid, crystalline structure of uniform strength and great hardness. Only a few elements even form such strong, tight, tetrahedral bonds.

Methane is the simplest organic compound based on carbon. It is a hydrocarbon; each carbon atom spokes out to 4 hydrogen atoms at a 109.5° angle. Its tension lines form a tetrahedral shape, but the hydrogen positions are not so tightly locked as in diamond, so methane is called a "plastic crystal."

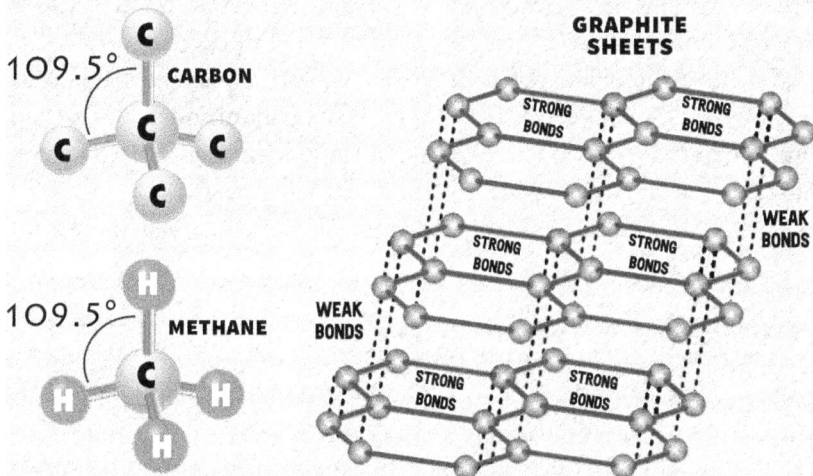

A diamond molecule—2 graphite sheets—a methane molecule

Graphite is chemically identical to diamond, but its carbon atoms do not bond into strong tetrahedrons. Instead, they bond in a planar ring system that stacks in graphite layers. The graphite molecules extend in two dimensions to form a hexagonal "chicken-wire" array of horizontal layers with weak electron bonds *between* the graphite layers. Such weak electron bonds between layers let graphite's fragile, slippery carbon sheets break or slide upon each other.

Science thinks that tetrahedral molecules are the basis for all life forms in this universe. Why? They allow for chirality in many organic molecules, plus a variety of geometries, and by extrapolation, lots of macromolecular versatility.

In the growing cosmegg's future, using tetrahedral bonds to generate organic life awaits far ahead. Now the cosmegg is still developing dimensionality, trying to solve space and time's *either/or* dilemma about tetrahedral volume.

But heed that tetrahedral shape! No wonder carbon took a cue from it! Tetrahedrons will be crucial in building the dimensional latticing of space and time, and those sturdy bonds will use all the polarities in the 64 hexagrams.

7. The power of the p-tree

Which primal, space or time, gets the 3D volume, and which primal gets shafted? The answer is simple enough. Clues to it already exist in many sciences and cultures. The master code has two lesser variants that people have studied thoroughly, so let's examine them both to figure out how the two latter-day variants of the original master code moved on beyond the polarized pair of pairs.

I Ching, genetic code, and master code...all three variants use the p-tree to grow their organizational systems. In Rosetta-stone style, we'll examine all three systems on the p-tree. By reverse-engineering the well-known structures of the I Ching and genetic code, we can gain insight into how the master code leaped beyond its polarized pair of pairs to universal success.

First Question: How does each variant establish its polarized pair of pairs on the p-tree? It's easy enough to see for the I Ching shorthand. Yin — — and yang ——— at the first level of bifurcation then fork again at the second level into stable yin ☰ ☰, changing yin ☰☰, changing yang ☰☰, and stable yang ☰☰. That's how the I Ching variant establishes its polarized pair of pairs.

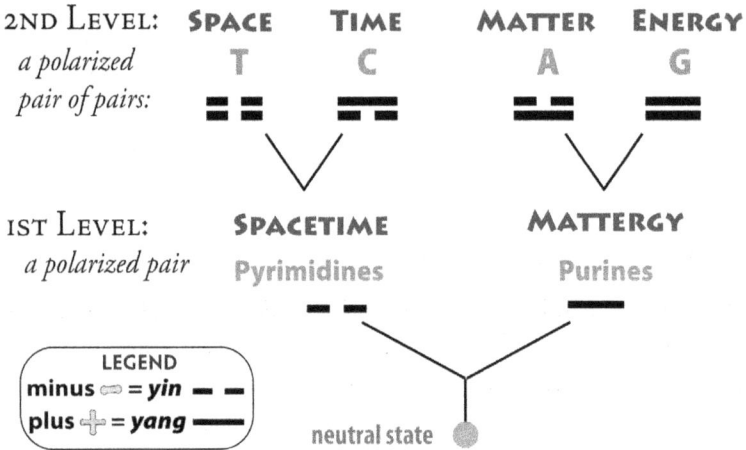

2ND LEVEL:	SPACE	TIME	MATTER	ENERGY
a polarized pair of pairs:	T	C	A	G

1ST LEVEL: **SPACETIME**   **MATTERGY**
a polarized pair   Pyrimidines   Purines

LEGEND
minus ⊝ = **yin** — —
plus ✛ = **yang** ———

neutral state ●

4 polarized paths = 4 bigrams, 4 molecules, & 4 primals

It's also easy to see how it happens for the genetic code. DNA's pyrimidines and purines at the first level of bifurcation then fork again at the second level into **T**hymine, **C**ytosine, **A**denine, and **G**uanine. That's how the genetic variant establishes its polarized pair of pairs. Book 3, *Tao of Life*, covers this in detail.

It's also easy to see how spacetime and mattergy at the first level of bifurcation then fork again at the second level into space, time, matter, and energy. That's how the universal variant establishes its polarized pair of pairs.

Second Question: "How do the two known codes handle their next stage of development?" That's also easy enough to see in the two lesser systems. Each variant leaps from the polarized pair of pairs into becoming pairs of polarized triplets. DNA pair-bonds its molecular triplets into 64 polarized 6-packs on the double helix. The I Ching pair-bonds its trigrams into 64 polarized hexagrams on a chart. Thus both lesser but better-known variants of the master code pair-bond 8 × 8 triplets across two domains to generate 64 polarized 6-packs.

How does the master code make the leap to higher dimensionality? We'll find out in Chapter 11. But as food for thought, Chapter 10 reviews how dimensionality has already developed at the mobic scale from a 0DD point—to a 1DD line—to a 2DD triangle. Now we're wrestling with a tetrahedron. Its 4 polarized 2DD triangles bond by merging their complex charges into 4 bigrams.

CO-SIMPLEX GEOMETRY'S
SPACE-TIME DIMENSIONS

Who gets to claim the tetrahedral volume...
Is it 3D space...
...or is it 3D time?

How many dimensions for this co-simplex?
tetrahedron ?

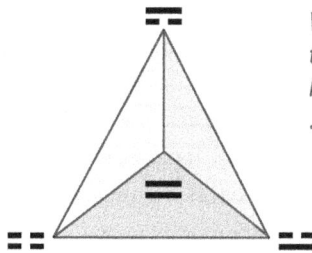

MOBIC-SCALE LIMITS

2DD co-simplex
plane

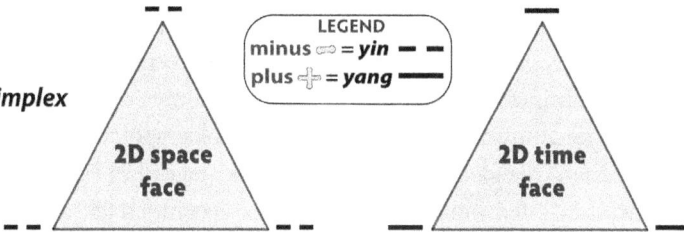

LEGEND
minus ⌐⊃ = **yin** − −
plus ⊹ = **yang** ──

2D space face

2D time face

1DD co-simplex
line

0DD co-simplex
point

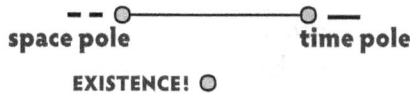

space pole   time pole

EXISTENCE! ○

So far in co-simplex geometry's space-time development...

Thus far, we've seen how all of the tetrahedron's planes fit together. So much for the polarity of their surfaces. None of that resolves the issue of tetrahedral *volume*. (Although clues are hidden in the chart above.)

The tetrahedron is being held together by polarized pulses of sheer being that subside back into nonbeing, continually reminding the cosmegg of the danger of ceasing to exist at all. What will resolve the cosmegg's dilemma so thoroughly that it jump-starts our enormous universe?

Buckminster Fuller thought the tetrahedron was the strongest structure possible. He said, "Four triangles can work together as a tetrahedral truss." But

Fuller extolled tetrahedrons whose planes are polarized by *space*.

In the cosmegg's ultra-tiny tetrahedron constantly forming and reforming at the mobic scale, each triangle has 2 faces. There are 4 triangles. They have 8 faces. Luckily, 2 + 2 + 2 + 2 + 2 + 2 = 8. And 2 doubled and doubled again = 8. And 2^3 = 8. All three operations take 2 to the same shared goal of 4 and then 8. Because all three 2→8 number paths cooperate so uniquely, their analinear congruence is a big factor in allowing our universe to exist. No other numbers can boast such a tight internal alliance as the 2, 4, and 8 possess.

Another factor is this: the tetrahedron has a trait unlike any other shape. If you take a tetrahedron and put its faces where its points are, and vice versa, you do not wind up with a shape that's just like it, oriented the same way.

No, instead, you wind up with a shape that's just like it, true, yet now it is oriented in a *mirroring* way that faces the opposite direction. In math-speak, "The tetrahedron is the *only* Platonic solid that does not map to itself by point inversion"…meaning only a tetrahedron can flip-flop its orientation in a mirror-reversing way. Since this unique mirror symmetry is possible only for the tetrahedron, it is another factor that allows our universe to exist.

The mobic-scale tetrahedron made of 2DD triangles contains more clout than an ordinary tetrahedron here in our upper bubble's 3D space and arrow of ½D time. Forced by dire straits to do so, the cosmegg will come up with something distinctly new. Something that is neither quite planar nor quite tetrahedral *per se*. Instead, it will learn to project holographic dimensional volume.

Since each triangle in the tetrahedron has a face of 2D space and a face of 2D time, the dimensionality of each triangle acts much like a laser beam that has been split into what holography calls a "reference beam" and a "reflective beam." On each 2DD triangle, its two faces already share and compare their space and time views via the constant polarized path of tension around them. They already act as each other's reference beam on dimensionality itself.

Now, dimensionality leaps to a higher order as the tetrahedral shape permits each 2DD triangle's two faces to meet and compare its own space and time layout with those of the other three 2DD triangles via the tension running on a path around its tetrahedral corners, like a baseball player hitting home run after home run. The scope of this new, grander view of dimensionality promises the score of a total win by providing enough foresight and foreplace for the cosmegg to construct the hologram of the Double Bubble universe.

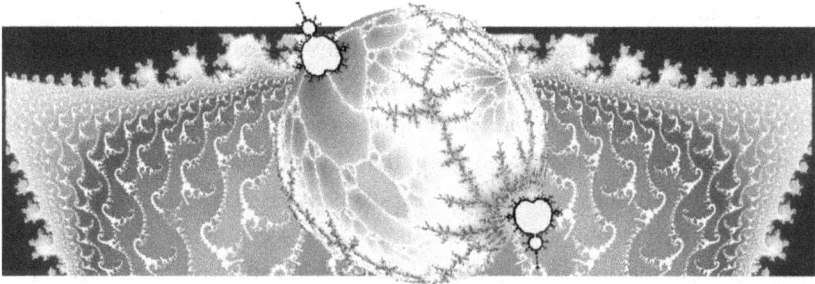

Chapter 10: Of Elephants and Carthage

1. George and the elephants

George Sudarshan, a theoretical physicist at the University of Texas, and his wife Bhamathi, also a physicist, attended for 8 years a dream group that I lead in Austin, Texas. A few times in that group, I mentioned some aspect of this TOE. Once when I spoke of tachyons, George perked up his ears and said, "I was the first to propose that meta-particle, saying it is faster than light." That perked up *my* ears, but when I asked after the session if he and Bhamathi would talk with me about the universal architecture I'd seen in my dream, they were not interested. George said he doubted a dream could say much about physics.

A group member who was watching this discussion later asked me why neither George nor his wife would discuss physics with me. I answered, "I don't know." Another member suggested that maybe George felt slighted by his own lack of recognition in physics. He directed me to George's *Wikipedia* entry, where I read that George had been passed over twice for a Nobel prize in physics on work he'd done. I also found online a PDF about George's work with Olexa-Myron Bilaniuk on theoretical faster-than-light particles…tachyons.

I wondered if maybe George felt no inclination to explore my dream's unorthodox perspective on physics because even his own rigorous work on more accepted topics had not received what he felt was adequate recognition. But of course, the Sudarshans both had busy lives, interests, agendas. Despite their lack of encouragement, I was nevertheless determined to keep exploring my strange, powerful dream until I could write down at least some part of what I'd seen in it well enough for others to see it, maybe even explore it on their own.

George did want to hear, though, about what elephants think. He became interested when I told the group one Sunday that my deep-see diving to explore physics also let me tune into mind talk with animal consciousness, particularly of the higher minds such as the primates, elephants, and octopi. I said such animals intuit more about this TOE than I do, that they access the lower bubble's cloudbank of data far more frequently than humans do these days.

Basically, the universal master code is so simple, yet so foundational, that at an intuitive level, it has been recognized many times by various members of various species throughout this universe. Here on planet Earth, it is intuited by some other species with higher consciousness…not through science, mind you, but in a mute, wordless knowing that some would call remote viewing. Robert Heinlein called it *grokking*. I experience it as deep-see diving.

I know that elephants and orangutans, even some of the few remaining huge octopi in the deepest ocean—what cold, clean, and wondrous minds they have!—recognize the cloudbank's meta-level of reality. However, we humans do not even view such creatures as being intelligent. We usually assume that intelligence means verbal skills and tool-making…just like us, in other words.

What those profoundly sentient beings know about the meta-level is so wordless that it exists in patterns organized below normal human awareness. They intuit a panoply of inchoate imagery coalescing into patterning that is both mental and emotional, resonating with a foundational reassurance. It shows them how the universe supports life even beyond a need for conscious thought. They sense the primal right structure of things so deeply that it offers them a certain meaning to life itself, a sort of perennial philosophy that says the universe grew them and wants them to be whole in themselves.

These animals know this, but they have no means to speak of it to us, nor to enforce their surmises upon us, so we consider them *dumb*—conflating somehow the word's two meanings of *voiceless* and *stupid*—and we assume that such animals are not just inarticulate, but far stupider than us.

We suffer as a result, we humans. We have lost some of our best allies by degrading some rather wise animals into the role of stupid, dumb beasts, worthy only to amuse us, carry our loads, do battle, or be eaten. I have mind-talked with some elephants in the shifting drifts of reality, and I find them often wiser than humans. Not more clever or sharp, but wiser. We humans are so clever and sharp that we often cut ourselves.

It is a shame that our species is so blind to elephant wisdom. If we had tapped into it millennia ago, we'd be further along in our quest for the secret of gravitation, a cure for war, and symbiotic harmony with this earth. We'd have re-examined why we do not respect the rights of other species that are part of this global organism of Gaia. We would know, for instance, that elephants have suffered imprisonment, lived through holocausts, undergone a slavery more systematic and brutal than any downtrodden race of humans has ever experienced. They are routinely killed just for two tusks of ivory and some teeth, sold away from their families, cut off from their culture and its teachings, consigned to cages, or set to work as stupefied slaves reduced to endless labor.

Image from St. Joseph's National School, Killala

Elephants do not have a written language, so I cannot exhibit any of their literary classics to you in a library. But if you heard elephants ponder life as I have, you would know that they are wiser in some ways than you and me, wise enough to have put a kinder servitude upon us than we did on them…if only they'd possessed the hands and guns and will to impose it upon us as we did on them. As it is, they may become extinct in the wild. But of course, we humans may become extinct, too, as self-canceling rabid killers who betray, poison, and fight each other and the Earth itself just for the hell of it.

2. Carthage and elephants

No wonder the media symbolize elephants as mighty and thrilling, a fit ride for monarchs. Or as innocent, cuddly babies in children's books. Elephants could have carried us much farther if we'd learned to access their wisdom in mind-talk instead of subjugating their strength and poaching their ivory.

You don't believe me? Around 1500 BCE, Phoenician traders moved beyond barter to the invention of money. They established Carthage in about 800 BCE on the north tip of Africa, near what is modern Tunis. By 650 BCE, Carthage was a commercial center of the western Mediterranean. Within another 250 years, it was rich and dominant due to its excellent harbor, the convenience of its money, and freedom to trade without outside interference.

To build various projects and transport goods, the Carthagenians used a subspecies of African Forest elephants from the Atlas Mountains. At that time, many such elephants lived in the mountains of Algeria and Morocco before drought and humans killed them all off. They were smaller, smarter, and milder in temperament than their bigger African Savannah cousins. They were also far more amenable to human interaction than were the large Savannah and Indian elephants. Atlas elephants had a natural philosophical bent so adept at tapping into universal mind that history might have gone quite differently.

If early Carthagenians had tuned into that species' willing, mild, meditative mindset, they could have gained much wisdom. Carthage might even have built a human/elephant rapport that could absorb both Greece and Rome before those two societies became a military threat and thus changed the course of history. Carthage could possibly have outshone Greece and squelched Rome.

If Carthage had conquered the Mediterranean basin by commerce and culture, then Hannibal the Great need never have taken elephants over the Alps in a desperate military strike against Rome. But that deeper synergy did not transpire between humans and Atlas elephants, so the road not taken became the road lost. I learned about that by deep-see diving into the cloudy currents of universal mind to examine alternate realities that could have been.

Other things might have changed, too. Elephants have more natural aptitude for abstract math than most humans do, but it is of an analog kind that we humans mostly only compute unconsciously, as do some autistic savants of music and math. Elephants can enter a state where they experience the foundational dynamic of nature's support without, alas, being able to articulate it to us or to themselves. But they still draw great comfort from it.

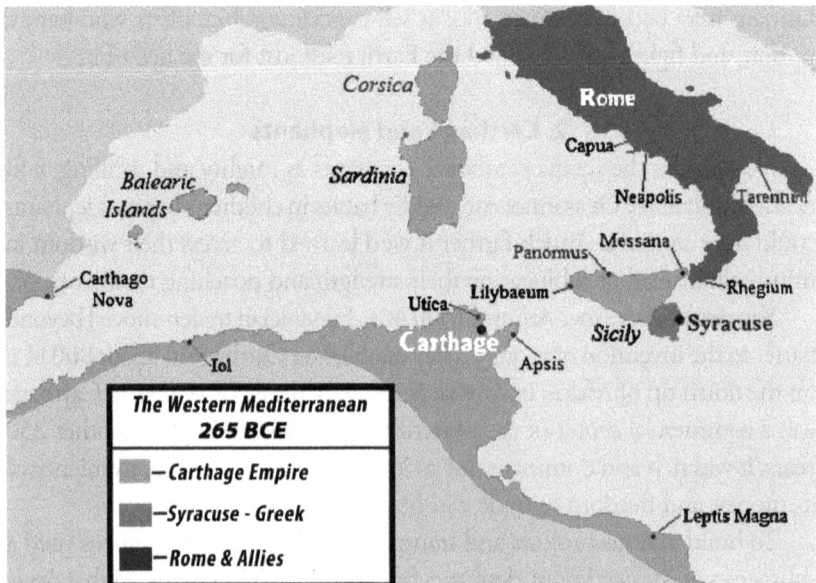

Carthage in 265 BCE

As it was, Carthage's commercial power did manage to keep nearby Greek city-states fairly well at bay. But Rome, that rising threat, greedily eyed Carthage's land and wealth, so the Punic Wars began. Hannibal's father, Hamilcar, was the Carthaginian lead commander in the First Punic War (264 to 241 BCE). He viewed the Atlas elephants much like military tanks, and he claimed that he lost the First Punic War by not having enough bigger Indian elephants at his disposal. The smaller, mild-mannered Atlas breed did not want to fight without great provocation. Too contemplative, too smart.

Hamilcar taught his son Hannibal military science, emphasizing the need for a large herd of big elephants in conducting battle. When Hannibal's father drowned on the back of an elephant during battle, the son decided to avenge his father's death by regaining all that territory lost to Rome.

In 218 BCE, Hannibal used elephants to march against Rome in the Second Punic War. He tried to sneak an army of 46,000 soldiers and 37 elephants over the Alps and the Apennines to surprise the Romans. That journey turned out to be far harder than he'd expected. About half of his men and a majority of the elephants died in that difficult trip. But Hannibal did succeed in bringing war into Roman territory; over the next 15 years, his Carthaginian army waged fierce battles throughout Italy. Many Roman and Carthaginian families lost multiple members in that warfare, yet neither side was decisively triumphant.

In 203 BCE, a showdown finally came at Zama, near Carthage. Hannibal had a huge army, plus a herd of 80 elephants, most of them Atlas. But when those elephants saw the massed Roman legions and heard their trumpets, they turned tail and fled, routing much of Hannibal's own army in the stampede. Carthage then fell to Rome and became a minor backwater in world events.

That rout reduced Hannibal's status so much that he became a vagabond for 20 years, and he finally suicided by poison. What an ignominious end for a soldier who'd fought so long and hard, humbled by not listening to the wisdom of elephants who were finally reduced to voting with their feet.

3. Emotional entanglement

It is no wonder that Hindu children dream of the elephant god Ganesh as a bringer of good fortune, that French children know Babar as an elephant king with human traits, that Disney-fed children see Dumbo fly with his ears. Elephants do soar upon what they hear below sound with a meaning that resides deeper than words. If we humans nowadays could hear it, too, we would know that elephants are philosophical.

Our dwindling herds of wild elephants meditate within the universal mind when they can turn to something beyond survival issues. Likewise for the threatened, reclusive, forest orangutans, and for a few giant squid still at great ocean depths, hidden as far away as they can get from the human marauders who are savaging and poisoning this planet. They still reach intuitively into the foundational wisdom, although most humans have lost conscious touch with it.

Those animals can still enter the lower bubble's resonant tachyonic cloudbank in meditation to grok a truth without words at the wellspring of reality. It is a great emotional comfort to them, the stability of this fundament...it's not just a shiny idea to turn about and tinker with as an intellectual toy. This deep level holds a shrine in nature itself. It honors the divine power that cares for us beyond care. Such animals fathom it with an insight below words, as we humans once did before we got so caught up in linear logic's left-brain glorification that we lost sight of that wordless, resonant reality, steady and true, existing below our bickerings and lawsuits, our accusations and frets about meaninglessness.

Even in captivity, some elephants can tune into the vast mind that loves them in the lower bubble. They do it for mental solace as they pace in their zoo enclosures and live only half as long as elephants normally do in the wild. They ponder and ache for their lost clans and culture and autonomy. They drag logs with a harness rubbing ulcerated sores on their hides, or they stand staked, teased, and abused by carnies bored for something to do. A few elephants become crazed by the futility of their circumstances, too insane to retreat into meditation or philosophy anymore, and that shallowness foments disaster.

In fact, I talk less and less to zoo animals these days. I am getting old and tired, I cannot help most of them in their enervating confines, and I grow weary of having to admit that to them. Mostly I just sit and accompany them in silence like another of their kind, puzzled by the implacable inhumanity of humans.

I wonder if you can accept the possibility that what I am telling you is true? Can it actually be true? Perhaps you'll consider it just an amusing fantasy...or a delusion...or maybe science fiction? Or just really stupid?

Let me approach this issue of mind-talk with animals from a different angle. In my experience, mind-talk at a distance occurs through both parties' willing agreement to do so. Both choose to be on the same wavelength. It cannot be done with arrogance, ill will, or aggression. Otherwise, you get tuned out. Agreeing to communicate brings you into the implicate order, which communicates among its parts all the time...just like your body does.

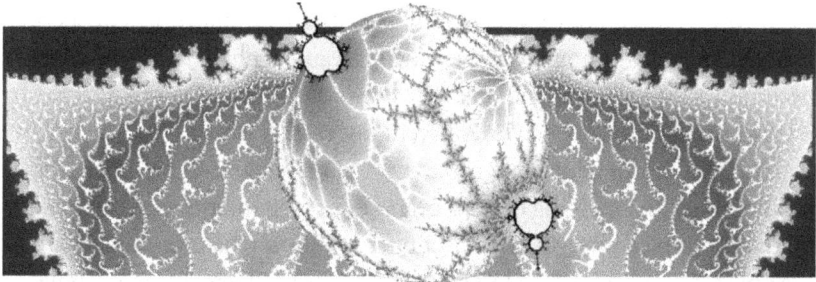

Chapter 11: Quest for a Holographic Universe

1. What is a hologram?

Our universe is alive, and it is holographic. Do those two traits sound contradictory to you? Or paradoxical? How about this? You are alive, and you are holographic. So am I. Do you protest that claiming we're living organisms and also holograms is antithetical? Impossible? Let me show you what I mean.

By every standard that humanity has put in place, you are alive. Agreed? If you are right now following what I am saying on this subject, and you personally want to agree or disagree or qualify your response, but you realize that I just won't stop spouting words, it suggests that you are alive. Agreed?

Maybe you insist that, okay, you are alive, but you are *not* holographic! You contend that a hologram is just a 3D image made of light. Photons. It's only an image, a faked 3D copy of the real thing, not the real thing itself.

Yes, that is true for the light holography we humans create here in our upper bubble, whether it be a *reflection* or a *transmission* hologram. And yes, holograms made of photons are not real 3D objects, only real-looking images made of shaped light. We humans can put a *reflection* hologram on a credit card. We can make a *transmission* hologram by splitting a unified laser beam into two beams, sending them on two different journeys, and then reuniting both beams at an angle that creates a 3D image of what they met on the trip.

How to make a transmission hologram

In the graphic above, a coherent laser beam is split into two beams. (*Coherent* means its photon waves are identical and in phase.) Those two beams become a *reference* beam and an *object* beam. Both beams get re-directed by 2 lenses and 4 mirrors to go on different but related trips.

The two beams eventually meet again at an angle to each other as they hit the apple. The merging of both beams creates an interference pattern that holds a scrambled image. Unscrambling it will show whatever scene or article the object beam was exposed to during the journey. Here it was an apple.

Since this graphic says the apple's scrambled image is captured on **Film**, you'd need a 3D projector to unscramble it. But if it were captured on a photographic plate, you'd need to focus a laser beam on that plate at just the right angle to unscramble the data and restore the apple's 3D image.

We usually realize that a hologram is not the real thing, but just a real-looking image. On YouTube, a holographic Carrie Underwood sings onstage with live Brad Paisley. Yes, Carrie looks "real." But she's not. Your eyes and ears say she's singing at the concert with live Brad, but her caption says, "Hologram." And, face it, you're even not seeing Brad in the flesh, either.

You can view a hologram even more simply by picking up a View-Master with disks. While using it, your two eyes act like the two split beams. Each eye sees a different but closely related picture of a scene taken from two slightly different angles. Your clever brain transcends those two images by turning the visual interference pattern into one 3D image…like your eyes are doing all the time anyway in ordinary life. It's an everyday miracle!

View-Master stereoscope and viewing disks

Back in 1952, my sixth-grade teacher, Mrs. Corinne Patterson, revealed to me the holographic miracle. She kept a stereoscope in a drawer of her desk, and occasionally during a geography lesson, we'd get to look through the stereoscope while she talked about some nation or national park. The students would pass around the stereoscope, and we'd all look through it. We could see the Eiffel Tower in Paris or the Grand Canyon (in black and white), and I'd pretend I was there. By my kid standard back then, those images seemed so real!

Mrs. Patterson pointed out that each round stereo card had pairs of flat,

2D pictures on it, taken from slightly different angles. When our eyes saw a scene from two different angles, our brains would put them together into one 3D image (black and white), much as our eyes did when we viewed the world around us (in living color). Here was explainable magic of the best kind.

2. Some background on the universal hologram concept

String theory and black hole theory have caused some scientists to suspect that we live in a holographic universe. They propose this universe may be made of 2D data projected like a 3D movie onto the universal horizon.

Where does that 2D data come from? Steven Hawking helped science realize how black holes play a key part in the concept of universal holography. He said if you want to reduce the universe of 3D reality onto a 2D surface, a black hole is a good model for it. Hawking even suggested that black holes are probably not really holes; instead, they are two-dimensionally flat and projecting a 3D illusion. He said maybe a black hole application to string theory could explain part of the universal holography process.

Charles Thorn pointed out that string theory describes a lower-dimensional reality (such as a black hole) where gravity emerges in a way that we consider holographic. But he could not prove it experimentally.

Then in 1993, Gerard 't Hooft proposed a holographic model of the universe based on the premise of a black hole's ability to store information.

Scientists cannot give precise numbers for holographic reality because there is no 0-point location from which to measure a true distance to the universal horizon, and there's no agreed-upon standard of measurement. Most physicists also think the distance keeps ever-growing, too, since they believe the universe is expanding constantly. They think the red-shifting light from other stars means that every part of our universe is speeding away from us.

In 1997, Juan Maldacena proposed a "holographic duality" of 5D anti-de Sitter space-time where the phenomena are entirely encoded in the behavior of certain quantum, non-gravitational fields taking place on the 4D border.

In 2003, Leonard Susskind resolved the holographic model into a more precise string-theory interpretation by combining his own ideas with those of other scientists, primarily Charles Thorn and Gerard 't Hooft. But again, it was only theory, not proved in a lab.

In 2009, physicists at a gravitational wave detector in Germany reported that they'd been trying to understand for years why their equipment picked up an inexplicable noise during their efforts to measure gravitational waves. (Personally, I would instead say they were measuring *holographic* waves, since in this TOE, polarity originates at the mobic scale and it emerges as gravitation when matter and energy appear at the quantum scale.

Craig Hogan, a physicist at the Fermilab particle physics lab in Illinois, thought the German team had finally "stumbled upon the fundamental limit of space-time—the point where space-time stops behaving like the smooth continuum that Einstein described and instead dissolves into 'grains,' just as a newspaper photograph dissolves into dots as you zoom in."

Hogan said, "It looks like GEO600 [the gravitational wave detector] is being buffeted by the microscopic quantum convulsions of space-time. If the GEO600 result is what I suspect it is, then we are all living in a giant cosmic hologram." But yet again, it was only theory, no real proof.

Yoshifumi Hyakutake and colleagues at Ibaraki University, Japan, tried develop a way to get proof. Between 2013 and 2015, they used computer calculations to make a holographic description of a quantum black hole. Their computations matched up with the 11D universe already suggested by string theory. Again, their work did not prove that the universe is a hologram, but at least it suggested the possibility that someday, computer calculations may be devised to calculate the quantum holographic properties of our universal bubble.

Victoria Jaggard described the idea in a 2014 *Smithsonian* magazine article, *What Is the Universe? Real Physics Has Some Mind-Bending Answers:* "According to quantum theories, if you examine the fabric of space-time close enough, it should be made of teeny-tiny grains of information, each a hundred billion billion times smaller than a proton…mathematically speaking, the fabric should be a 2D surface, and the grains should act like the dots in a vast cosmic image, defining the "resolution" of our 3D universe…these grains should experience random jitters that might occasionally blur the projection and thus be detectable." Again, I would instead call those jitters *holographic*.

Of black holes, Jaggard said, "Black holes are born when dense packets of matter collapse in on themselves…. Some versions of the equations…say that the compressed matter does not fully collapse into a point—or singularity—but instead bounces back, spewing out hot, scrambled matter…everything in and around us would be made from the cooled, rearranged components of that scrambled matter."

A 2017 study in *Physical Review Letters* reported that Canadian, Italian, and United Kingdom scientists together found substantial evidence to support a holographic explanation of the universe, in what they "believe is the first observational evidence that our universe could be a vast and complex hologram." Irregularities found in the cosmic microwave background suggested to them that there's at least as much evidence to support a holographic model of the universe as already exists for the current inflationary model of the universe.

Physicist Brian Greene summed up the concept in *Geek's Guide to the Galaxy*, podcast episode 60: "The idea is we may be that 3-dimensional image of this more fundamental information on the 2D surface that surrounds us."

In other words, all of us observers act like View-Masters. Our eyes unscramble the universe's 2D data into this seemingly-solid 3D reality.

But wait! If I close one eye, do I see scrambled data while my other eye is shut? No! Not only my vision, but instead, all of my senses insist this world is real. They all declare our reality is more than just photons making images.

3. The universal hologram is physical, mental, and spiritual

The modern mindset usually considers a holographic universe only scientifically. That narrowed view has led some people to decide we are living in a matrix that is holding us prisoner. Some even insist we must find a way to escape it. Others, stunned by "virtual cleverness," have proposed that we live in a high-quality game that is being played by a much smarter alien race, entertaining itself by tweaking us around, so we are without any free will and merely imagining that we can make decisions to affect our lives.

I suspect such dystopic constructs come from a techno-infected way of looking at life and reality. The idea of a light-based reality is quite ancient, actually, but back then, people did not call it "holographic." Plato, for instance, spoke of reality as watching shadows play on a cave wall. Hinduism said the god Vishnu is dreaming the universe, which exists until he awakes. The Amazon basin's Uitoto people say their god Nainema, himself an illusion, ponders illusion itself, rather like two angled beams generating the world from nothing.

Michael Talbot's most famous book on parallels between ancient mysticism and modern science was *The Holographic Universe*. It described in great detail how ancient cultures experienced holographic reality in a time when people mostly did not look at nature scientifically. They just experienced its wholeness.

Most ancient cultures believed the universe is one holistic unity experienced in three aspects: physical, mental, and spiritual. In modern times, an interesting approach comes from Hawaiian professor Manulani Aluli Meyer. She writes and lectures on the holographic epistemology of indigenous concepts. For instance, in a University of Hawaii lecture in 2011, she considered "Holographic Epistemology: Native Common Sense."

Several current groups of scientists, mathematicians, and philosophers have joined in study projects to consider the holographic, holistic nature of the universe from multiple humanistic perspectives. One such current group offers the online 5-part series called *Butterflies are Free to Fly*.

4. The universe is a deeper, more basic kind of hologram

This TOE says our holographic universe started smaller than current science can measure. It didn't originate at the tiny quantum scale where matter and energy emerge. Instead, it originated at the ultra-tiny mobic scale where space and time emerge. The process started when pulsing *being* versus *nonbeing* (0/1) set an *on*-dot of sheer existence…it's a cosmegg! Then two consecutive *on*-pulses describe a polarized line of dimensionality with two poles—space and time.

SPACE POLE -1⊶⎯⎯⎯⎯⎯⎯⎯⎯⊷+1 TIME POLE

But to get a hologram going, the cosmegg needs more ways to complicate its original line of space-time tension. By setting just one more point in a new location, dimensionality gains new clout, testing the boundaries of that ultra-tiny scale by describing a 2DD triangle with two polarized faces exploring two different but related dimensionalities: a 2D space face and a 2D time face.

CO-SIMPLEX GEOMETRY'S SPACE-TIME DIMENSIONS
AT THE SCOPE SCALE

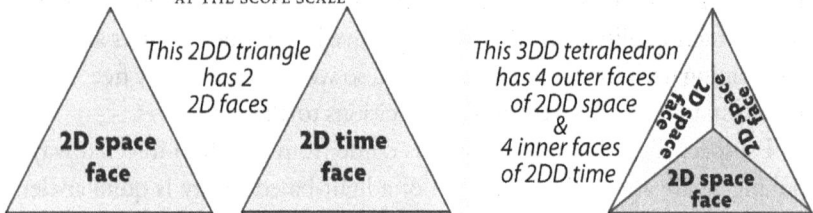

This 2DD triangle has 2 2D faces

2D space face

2D time face

This 3DD tetrahedron has 4 outer faces of 2DD space & 4 inner faces of 2DD time

2D space face
2D space face

2D space face

Four 2DD triangles form the surfaces of a tetrahedron

Dimensionality gets another boost when the cosmegg leaps into a still-higher order of organization. With just one more point, it becomes no longer a two-sided triangle. Instead, dimensionality has *4 different* two-sided triangles. They form a tetrahedron of dimensional tension. Those triangles flash constantly in and out of existence, transitory, yet able to bond with the previous and future triangles due to 2DD space-time's encoded disposition of foresight and foreplace.

Recall how the image of transmission holography at the front of this chapter showed 4 mirrors? The cosmegg's loosey-goosey tetrahedron now has 4 quick-flashing 2DD triangles, each with a 2D space face and a 2D time face. Its 4 triangles have 8 faces—4 faces of 2D space and 4 faces of 2D time. By our human standards, that doubles the number of dimensionally-reflective surfaces needed for holography, a doubling that harbingers the Double Bubble universe.

On every 2DD triangle, space and time sit in such tight alliance at this ultra-tiny scale that they always know what each other is doing. Such accord of spacing and timing among all 8 planar surfaces means every datum grasps just how it fits into the whole. Then, too, information doesn't have to transmit very far, just to the next adjacent triangle as the 2DD information ever-renews, constantly updating the view of reality on all four 2DD triangles.

5. Facing the challenge to go big

From our ride on the dimensional roller coaster, we watch the 2DD planes flash into existence and bond into a tiny, transparent, flickering tetrahedron at the mobic scale. To me, it recalls the Louvre Museum's glass pyramid…except the cosmegg's pyramid at the mobic scale has a triangular base, not a square base.

As its 2DD triangles constantly flash in and out of existence, we cannot determine the tetrahedron's 3D volume. Why not? Space and time have not yet resolved their volume quandary. Each still seeks to push its volume out beyond the mobic scale. Who will win the 3D volume prize? Space or time?

The cosmegg's dynamic drive is not conscious mentation. It is merely part of the living cosmegg's innate urge to achieve sustainability. All living things have this drive. For instance, the cosmegg's polarized pulses will eventually birth gravitational force at the mobic scale. Gravity makes things move. Water doesn't think about seeking the lowest level. It just does. Apples don't choose to fall. They just do.

With volume, the cosmegg can go analog (-1 and +1) and also go linear binary (0/1). Thus the notion that either space or time must win the 3D volume is a false, forced choice. It can go *analinear,* going big by choosing both!

How does the cosmegg make this leap to claim two volumes? Easily. Its tetrahedron is made of four 2DD triangles, so its volume is not just 3D, but rather, 3DD. This 3DD tetrahedron has *two* kinds of volume: an exterior volume of 3D space and an interior volume of 3D time. Moreover, both volumes *must* project, for they cannot manifest at the mobic scale, which can only hold rapidly flashing planes that coordinate data via their hindsight and foresight of the previous and upcoming 2DD triangles.

From our overview of emerging dimensionality, we realize that the war between space and time for tetrahedral volume is actually an opportunity for both. As we watch dimensionality's double-win manifest, let's make a video.

Video: the cosmegg counterbalances space and time by projecting *both* volume options of the 3DD tetrahedron, inner and outer: 3D space extends far above the mobic scale, and 3D time extends far below it. For utility, we'll depict this mirror-twin expansion as an hourglass-shaped cell ⧖. Its 3D space stretches far above the mobic scale to the universal limit, and its 3D time stretches far below it in similar fashion. This is a 3DD cdp hull.

News flash! *Any* tetrahedron has an inner and an outer volume. Always. But here in our time-poor upper bubble, if we consider an object's volume, we naturally assume its volume exists *inside* the object, not *outside* it. Oh sure, we might paint the outside *shape* of a 3D object. For instance, we can paint the outer walls of a house. But as we paint that house, we are actually painting only its outer 2D *surfaces*, not the outer 3D *volume*. A house's outer volume would technically include all the open space beyond it!

Moreover, we automatically assume that the word "volume" implies *space*, not *time*. It's a natural thought. Euclidian geometry did not require us to think about time having any volume at all…no need for such whimsical notions way up here riding on the arrow of ½D time above the mobic scale.

Here in the upper bubble, we experience the gracious volume of 3D space and the thin arrow of ubiquitous time, so we just take it for granted that volume obviously refers to space, and moreover, even to *interior* space. We also keep finding new ways to corral and divvy up that interior space. We parcel it into cups, cubic meters, oceans…and they all hold inner volume, not outer.

But such is not the case for the 3DD tetrahedron whose 2DD triangles constantly disappear and re-form at the ultra-tiny mobic scale. Right here, *volume* does not necessarily mean *interior,* nor does it necessarily mean *spatial.* It can be interior *or* exterior—and it can be temporal *or* spatial. In fact, at this ultra-tiny scale, it is mandatory that both volumes find a means to express themselves elsewhere. They cannot fit into the mobic scale. For both volumes, the 3DD cdp solution resolves the conflict between space and time.

CO-SIMPLEX GEOMETRY'S SPACE-TIME DIMENSIONS

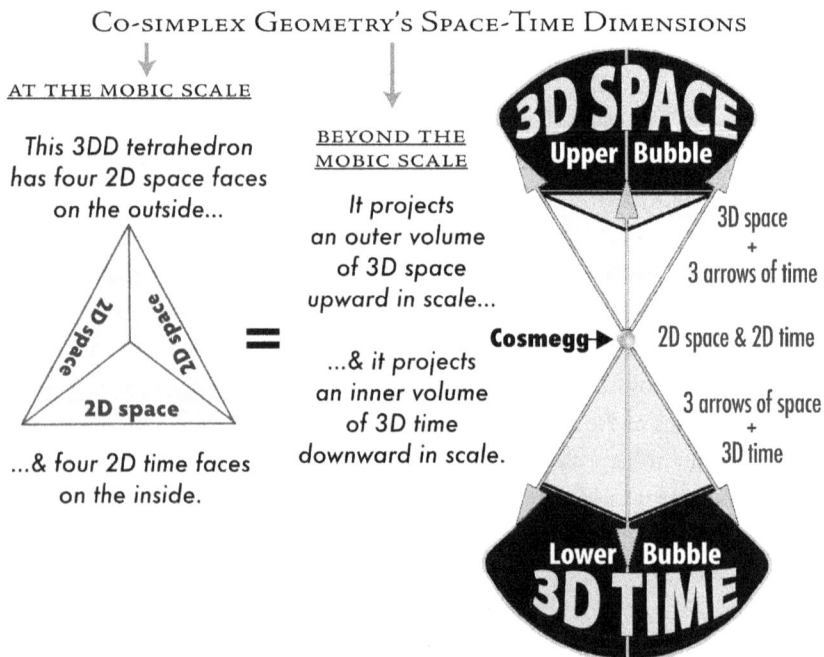

AT THE MOBIC SCALE

This 3DD tetrahedron has four 2D space faces on the outside...

BEYOND THE MOBIC SCALE

It projects an outer volume of 3D space upward in scale...

=

...& it projects an inner volume of 3D time downward in scale.

3D SPACE
Upper Bubble

3D space
+
3 arrows of time

Cosmegg

2D space & 2D time

3 arrows of space
+
3D time

Lower Bubble
3D TIME

2D space

...& four 2D time faces on the inside.

Two ways to symbolize a 3DD tetrahedron's 2 volumes

Now, from a single 3DD tetrahedron, the cosmegg has developed one hourglass cell projecting its two 3D volumes far out beyond the mobic scale. 3D space projects above the mobic scale, and 3D time projects below it. At the hourglass cell's wasp-waist sit the 4 tiny, flashing 2DD triangles developing the polarized bonds of a 3DD tetrahedron that is constantly iterating at the mobic scale and projecting its pair of 3D volumes outward beyond that scale.

Do you recall from earlier how a looping, polarized 8-path ran around both sides of a 2DD triangle? Now in the 3DD tetrahedron, all those polarized 8-twists get projected across both 3D volumes of its hourglass cell, becoming 3 constantly moving 8-loops that ride along the 3 corners of the hourglass cell.

The projected result looks sort of like two triangular megaphones joined at their mouthpieces. One megaphone holds the volume of 3D space with three stress arrows of ½D time running outward along its triangular corners and then looping back in toward the mobic scale. The other megaphone holds the volume of 3D time with three stress arrows of ½D space that likewise loop.

The upper half of an 8-loop reads as the arrow of time above the mobic scale, but its lower half reads as the arrow of space below the mobic scale. The 8-path repolarizes where its two megaphones meet. In other words, a moving 8-loop switches polarity on each pass through the mobic scale, turning from time into space, or from space back to time, whenever it crosses the interface.

By this means, dimensionality leaps to a still-higher order of organization that updates the Rule of 4. How so? Recall that down at the mobic scale, the perpetually re-emergent tetrahedron is made of four flashing 2DD triangles. They are a polarized pair of pairs, Further, each 2DD triangle has a 2D space face and a 2D time face. In this way, the tetrahedron's planar surfaces adhere to the Rule of 4. All of this echoes the rule of 4 and thus reinforces planar security.

However, the tetrahedron's 2 volumes, 3D space and 3D time, do not honor the Rule of 4. Instead, they now transcend it. The pair of 3D volumes in an hourglass cell are a polarized pair of *triplets*. It's a cellular 6-pack of dimensionality. It also holds 3 vectoring 8-loops made of 2 orbits each, giving the hourglass cell a second polarized pair of triplets, i.e., another 6-pack. This leap into 6-packs foreshadows DNA's 64 molecular 6-packs and the I Ching's 64 lineal 6-packs.

The hourglass cell's dimensional 6-pack of expanding volumes and its tensor 6-pack of arrowing orbits use analinear flex-math that gives the cosmegg enough dynamic stability to go ahead and generate the Double Bubble universe.

Instantly in the "nothing" around it, the cosmegg replicates that single hourglass cell into many hourglass cells. They all merge to form the holographic bubble of 3D space above and the holographic bubble of 3D time below. Its instantaneous cell proliferation and merger into the living body of our Double

Bubble universe is what humanity's upper-bubble science calls *inflation!*

Both bubbles have fractal traits iterating from their tetrahedral origin that will develop the latticing of 3D space in the upper bubble and the latticing of 3D time in the lower bubble. In each bubble, its 3D lattice bonds via 8 × 8 = 64 different 6-packs of polarized pulsing. The grid can scale up and down in dimensional size, and its bonding patterns can morph according to need.

The Double Bubble's latticing of 3D space and 3D time accommodates the tensor network of 8-loops vectoring across it. To be clear, the upper half of each 8-loop reads to us in this upper bubble as the arrow of time moving ahead in 3D space, whether it heads upscale or downscale, whether its vector is rising to the utmost limit or sinking back down toward the mobic-scale interface. "Why?" you may wonder? It's because the vector is always moving ahead. Likewise, the lower half of each 8-loop in that other bubble's 3D time always reads as the arrow of space, heading down or up. Why? Because it's always moving ahead.

As we'll see soon, that last leap into 3D construction puts an end to the universe's proliferating dimensionality and sets matter and energy into the chute of possibility. But the master code grown on a dp-tree becomes a template for future lesser variants of the paradigm within the Double Bubble universe.

The great organizational leap from a polarized pair of pairs to a polarized pair of triplets has let dimensional latticing bond via 8 × 8 = 64 6-packs of co-chaos patterning. The same dynamic will be utilized in lesser variants, including this bubble's genetic code that makes myriad little living organisms. DNA molecules float around in our cells, able to bond as polarized pairs of pairs, but then leap to a higher order as polarized pairs of triplets that bond into DNA spiraling along every double helix in 8 × 8 = 64 basic 6-packs of co-chaos patterning, and their bonding patterns can morph according to need.

The I Ching's ancient math shorthand is yet another variant templated off the master code. It can bond a polarized pair of pairs into 4 bigrams, but it can also bond 8 × 8 polarized trigrams into 64 different 6-packs of co-chaos patterning in 64 hexagrams that describe the way of the Tao, and their bond patterns can morph according to need. These evident variants and many other broken, thus less evident symmetries appear in our living, holographic universe.

You and I and this universe are alive, real, and holographic. But you and I and it are not made of shaped light. We are made of shaped space, time, matter, and energy…only one component of which is photons.

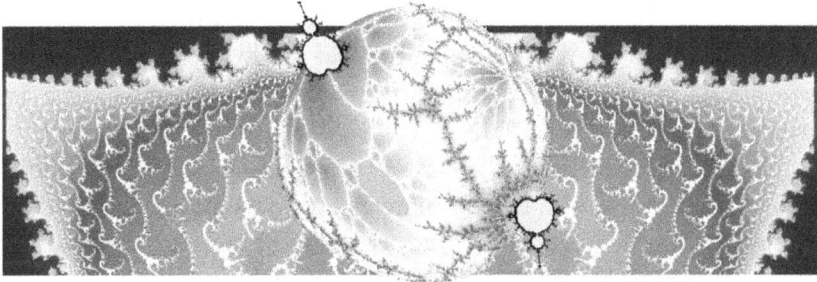

Chapter 12: Dimensional Clout

This chapter takes a quick look at dimensionality discussed so far in this series, both mathematically and philosophically. It considers the origin of dimensionality that we explored with a variant of simplex geometry. It also considers why we exist in this bubble and where that purpose aims.

1. Co-simplex geometry

Simplex geometry generalizes the idea of a triangle or a tetrahedron (made of triangles) to arbitrary dimensions in space. Triangles are the only shape it uses to explore dimensionality, so you might say this branch of geometry is fixated on triangles, as if they are the only shape that really matters.

In simplex geometry's space dimensions above the quantum scale…

- *The 0-simplex is a 0D point in space*
- *The 1-simplex is a 1D line in space*
- *The 2-simplex is a 2D triangle in space*…and from this stage onward, every simplex shape is made of triangles.
- *The 3-simplex is a 3D tetrahedron in space*

This TOE uses a variation of simplex geometry to describe dimensionality originating at the mobic scale. At that ultra-tiny scale, co-simplex geometry generalizes the idea of a triangle or tetrahedron to a non-arbitrary number of dimensions in both *space* and *time*. Thus this geometry deals not only with space dimensions but also with time dimensions.

Co-simplex geometry starts at the ultra-tiny mobic scale, where dimensionality begins with the 0-co-simplex of a 0DD point.

In co-simplex geometry's space and time dimensions at the mobic scale…

- *The 0-co-simplex is a 0DD point in space and time*
- *The 1-co-simplex is a 1DD line in space and time*
- *The 2-co-simplex is a 2DD triangle in space and time*
- *The 3-co-simplex is a 3DD tetrahedron in space and time*

At the mobic scale, the 0-point of nothing has no space or time dimensionality. Then a second point appears in elsewhere and elsewhen, establishing a line of dimensional tension between both points, which are polarized by space and time. This is a tension line of 1DD space-time.

Another point turns that 1DD line of tension running between two poles into a 2DD triangle with tension running around its 3 corners. The 2DD triangle has two faces of polarized dimensionality. Holographically speaking, those two faces act like two "beams" of 2D space and 2D time existing at the mobic scale. They act as each other's reference beam and object beam.

Then a final new point reconfigures the 2DD triangle into a 3DD tetrahedron. It is composed of four fast-flying 2DD triangles with eight triangular faces. The four space faces are positioned outward on the tetrahedron, and the four time faces are positioned inward on the tetrahedron. All those fast-moving faces can remain at the mobic scale, but the tetrahedron's two volumes cannot. So where do the two volumes go? The answer is in Chapter 13.

2. Moving around in space and time

In this white-hole bubble, life signs us up to experience events occurring in 3D space over ½D time. At birth, we pop into the playpen of 3D space and begin to move around. It is only later, at perhaps the age of 5, 6, or 7, that we start to abstract enough to realize our bodies are inhabiting three dimensions of space simultaneously, and we can even name them…length, width, and height.

However, from birth, the constrictive vise of time immediately grips us in its inexorable constraint of right *now!* The arrow of time propels us onward through feeding time, bath time, nap time, and bedtime…that awful crying moment when Mommy or Daddy disappears to leave us alone what seems like eternal *now* until we drift to sleep.

Wheedling may get us out of the crib, but nothing can fast-forward us enough to skip two hours of colic or a year of cancer. All lifelong, we ride continually penned on the arrow of time. Its moment of *now* is always flying toward the future. We are forced to ride shackled on time's line of flight.

Thus, here in this upper bubble, we can cavort in 3D space but live imprisoned on the arrow of time.

Space-wise, no complaints. There's plenty of room up here. We have leeway to stretch and move around in this gracious upper bubble. Light, air, and water exist up here because our 3D space permits matter and energy to expand enough to elaborate into atoms and even full-blown, expansive molecules that build our separate bodies with the freedom to frolic in all three dimensions of space at once. Our sensory organs provide data on how to maneuver in it. Our muscles operate on lever-like limbs that carry us across the distance between various material structures…maybe while shopping at s store, for instance.

Time-wise, one continuous complaint. We live pinned on the arrow of time, physically locked here…but our thoughts started learning to transcend the time prison. It was growing foresight about time that allowed us to notice the Nile's yearly rise and fall, decide to settle down, plant crops, and then harvest them in the progression of seasons throughout a cycling year.

With each new leap of consciousness, we kept fortifying ourselves against the vagaries of time by setting hedges against our own ephemerality. We've built reputations. Pyramids. Dynasties. By now, we have banks for money, food, seed.

3. All roads lead home to love

Meanwhile, in the lower bubble, a quick, unified, universal mind keeps selecting for more consciousness in the tiny, diverse mini-minds here in the upper bubble that it's been cultivating over eons like hot-house flowers.

That lower bubble's vast mind keeps track of events, but it does not really control the alternative possibilities. Why? Because the analinear paradigm allows many free-will choices for us myriad little organisms proliferating in this bubble. Our Double Bubble universe is designed to inform and evolve the Grand Organizing Design that is far bigger than this universe, or all of them put together. It uses the myriad minds in various universes to explore, diversify, and amplify the permutations of love as a means of connection to the divine. By living out our lives, we meet, greet, and refine the divine.

In all my exploring, I've found that love—love of all kinds—seems to be the underlying motivation for this universe's creation. We are here to delve into love and amplify it. This sentient universe probes for the awareness and testing of love in all its available parameters. Even in things that to us seem insensate.

For instance, to us, the sun does not appear sentient or even alive. However, the star that is our sun has a rudimentary sort of consciousness, and it loves. In deep-see diving, I've talked with it and a few other stars…but when I say *talk*, I do not mean that literally. Instead, it is a sort of mind-reading, or more exactly, an exchange of awareness at a diffused, generalized level.

Our sun's love is utterly self-intent, yet it is in no way egomaniacal, just merely simple. Meanwhile, it works for us. Fission explosions occur at every second in its hot core. Those constant explosions do not jeopardize the universal life plan. Instead, they foster it. The continual explosions in our sun exist in a fortuitous placing and timing that allows us fragile beings to exist here on Earth. As a source of energy, those perpetual explosions radiate heat and light to nourish us bits of organic life positioned just far enough away to thrive on this third rock out, orbiting around its hot body that warms our lives and lights our way.

And this sun warming this Earth is merely a small part of the universe's sorting of possibilities to foster consciousness over the long haul by developing myriad mini-organisms in climes wherever they can thrive.

Stars that I have sought out to communicate with seem to have some sort of integral awareness of their own being. Each has a discrete, individual sense of itself. Each exhibits an awareness of its own particular internal dynamics and even of its niche within its ilk. But stars, whether they are wreathed in planets or not, are not very aware at a mental level, not compared to us.

Stars just focus on a consuming love for their ongoing processes. They lack curiosity and had no patience for mine, either. Stars actually resent being taken away from focusing on the internal fires and forces and pressures of their physical makeup. Life becomes for them all-consuming, a rapt naval-gazing upon their own burping and belching and burning up with the fever of being.

Stars live in a sort of exulted yet myopic fixation on the holy fire of self-immolation that takes each one ever more outward or inward. They certainly do not have souls of ongoing existence beyond each star's unique lifetime. Sufficient unto themselves, yet affecting others constantly, each star is entirely satisfied to live and die upon the sword of its own light.

4. Testing the extent and limits of all kinds of love

In this upper bubble, we are all capsules of portable mind manifesting in matter and energy above the quantum scale. The universal mind operating in the lower bubble got the evolution of all us tiny mites in the upper bubble up and running. It started the thrust of evolution going for every species, and it keeps them aligned enough that they tend to support each other by their differences.

That vast mind on the other side of the mobic scale knows how this universe grew from a 0-seed of divine intent to explore the productive power of creation. Think of it: our Double Bubble is based on the bifurcation of nothing into the plus and minus of something. It predicates the energy of yang and yin moving into a third state that creates. This drive is inherent in all events up here. In each being, it tweaks the plot to aim toward the same place: knowing the unknown about productive love.

The unified mind in the lower bubble is trying to bring us to the point where we can recognize consciously, not just unconsciously, that it too lives. Not only that, even to recognize that it desires a more conscious relationship with us.

Our Double Bubble universe is based on a polarized bifurcation of the number 0 that grew by continually integrating those poles in level after level of rising complexity, evolving higher orders of polarized power. That's how each species emerged and evolved, how our drives, intentions, values, and ethics emerge and evolve, our hopes and regrets, too.

Why bother to set up so many species with their explorations and expressions of the nature and limits of love? Our maturing universe is tasked to investigate the creative power of every kind, version, and volume of love.

All of this creation accommodates the shape and thrust of a triumphant reality that not only survives but thrives, not only thrives but keeps refining its expressions of love. Why? Because God wants to know how to improve the power of love. Love in the upper bubble is playing a high-stakes card game, while the lower bubble's mastermind keeps tweaking all the hands to win.

In doing so, the Double Bubble fulfills its own great life by operating as a mathematician-cum-engineer-cum-architect-cum-artist cum-mystic, all intent on probing the divine directive. All creation is tuned again and again toward its goal of union with a creator larger than all the universes beyond this Double Bubble, with all of them enlisted in the Grad Organizing Design we humans inadequately call God.

So be assured that our Double Bubble universe has an intentional mind. Its universal momentum is predicated on bifurcating, polarized numbers that elaborate into co-chaos fractal patterns testing the extent and limits of all kinds of love. Where does it aim to go? Hopefully, it will continue to live long and well and evolve into something even greater and more wonderful. The options playing in the lattice of the upper bubble may someday resolve into the unity of one universal conclusion...if we finally get done with all this effort at calculating the value and meaning of love...if we get our conclusions into good shape...if we finally sum things up with love. Or not. We shall see.

5. The new incarnation of endless play

The evolving great patterns of resonant affinity and disaffinity could go on and on, theoretically speaking. But they likely won't. Now I am touching on ⟐ **Question 19: How will our universe end? Or will it?** Will our Double Bubble go out with a bang? A whimper? The scream of a big rip tearing everything apart? Will it dwindle into smooth, cold entropic soup? Or become a black hole? Or bounce in endless cycles of inflating boom and contracting bust?"

None of the above, says this TOE. The curtain comes down on our Double Bubble play whenever the universal mind as master playwright gets its point across—to itself, to us players, and to the grand audience of God. But our theater will not close down as a bleak, dark void expanding forever. Nor with a bang, nor a whimper, nor in a bouncing, self-repeater syndrome.

No, it will likely leave with an applauded bow to open again in a new, expanded, upscale venue. Due to possibilities arising from the integrative consciousness of this universe, it may cocoon up in about 5 billion years from now, clenching into something small and dense, rather like a firm, fat, foam, foursquare pillow (or like a solid agreement between itself and God made compact), and then it will balloon out again as something so huge and completely different that it becomes a universe of nearly endless delight, fully conscious, fully realized, a massive entity rolling in endless play with God.

This seems more and more like a strong possibility, for which our universe is pleased, looking on down the pike. That will make it a rare universe indeed. Such has never happened before in this particular way, I'm told. Told by whom?

By tapping in somewhere along the line of communication between the universe and the Grand Organizing Design. Do I get to see their tweets? Or instagrams? You see, I must rely on approximations for a communication line so far beyond words, images, apparatus, beyond even consciously formulated ideas. It is difficult for me to verbalize, but I know it like the great octopi know it. It is knowing at a cellular level where the perennial philosophy still resonates from back in a time when people lived closer to nature, and to divine force.

God, too, is pleased and excited about the likely upcoming massive universe of conscious play, I hear, hoping for increased communication, even for a heaven of the soul's delight. Heaven is what you make it, so it is better to keep the parameters open and leave the final specs up to God. Meanwhile, our Double Bubble universe hopes to meet God eventually in a rarefied companionship that only that new incarnation into a fully conscious, huge, and playful universe can provide. It would be a heaven worth working for, or so I hear.

Hey, this universal testing of love is not a dogma that I expect you to embrace or refuse. After all, how exact, how testable, how irrefutable is it? I cannot mathematically prove to you that love is the focus of research and development in this Double Bubble universe assigned to explore love through the evolution of consciousness in bits of matter and energy carried in their space and time containers.

I'm merely relating to you, as best I can, what I saw at depths that I cannot fathom, what I experienced in places that I cannot describe, so take it with your own grain of salt. Weird, wild…and maybe wonderful? Yes.

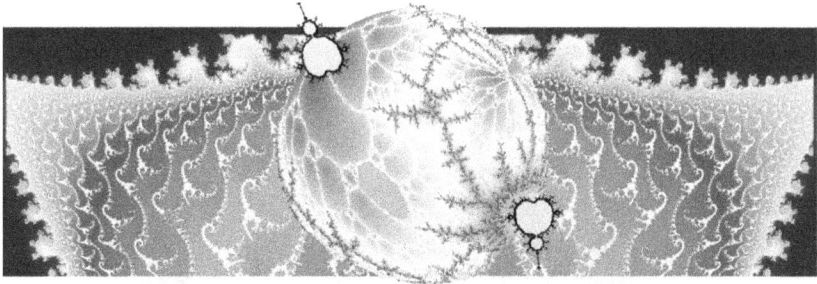

Chapter 13: From Tetrahedron to Tetrastar Cube

1. Ben Franklin guessed wrong about electricity's poles

Our dimensional roller-coaster ride lets us view gravitational pulses sparking along the 2DD tetrahedron's transparent planes from point to point. Is the continuous beat of each pulse sparking *positive* or *negative* on the tetrahedron?

Ben Franklin wondered something similar about a different force when he used a thunderstorm to snake electrical charge along the wet hemp string of his kite and electrify a dangling key. Before atomic theory, Franklin and many others assumed that an imbalance of "electrical fluid" caused a lightning bolt. They supposed its charge flowed like air pressure or water pressure, from a more heavy, positive pressure toward a lesser, negative pressure.

In the storm, Franklin wondered, "The electrical fluid flows how? From kite to lightning bolt? Or vice versa?" It was happening too incredibly fast for him to see or test. At home, too, when he did experiments to generate electrical charges by rubbing glass with wool or silk, Franklin again wondered, "The electrical fluid is flowing in which direction? From glass to wool? Or vice versa?"

World convention had long labeled *assertive, active, yang energy* as positive and *receptive, passive, yin energy* as negative. Franklin wanted to follow that convention, so he sought to label the pole assertively sending out electrical charge as positive + and to label the pole receptively accepting its charge as negative -. But he guessed wrong and named electricity's assertive pole negative - and its receptive pole positive +, the opposite of how electrical charge actually flows.

Franklin's topsy-turvy terms got baked into modern science. We still label the two electrical terminals on a battery the opposite of how their charge actually flows. This habitual, cognitive doublespeak in science decrees that on a battery, the terminal we call negative - actively sends out electrical charge, while the terminal that we call positive + passively receives it.

Another topsy-turvy example? Below, a *Wikipedia* drawing of a lightning strike shows a cloud sending down a long, assertive, active (red) leader that is paradoxically labeled negative -. The building's lightning rod has a short (blue) leader that receives and grounds the charge, paradoxically labeled positive +!

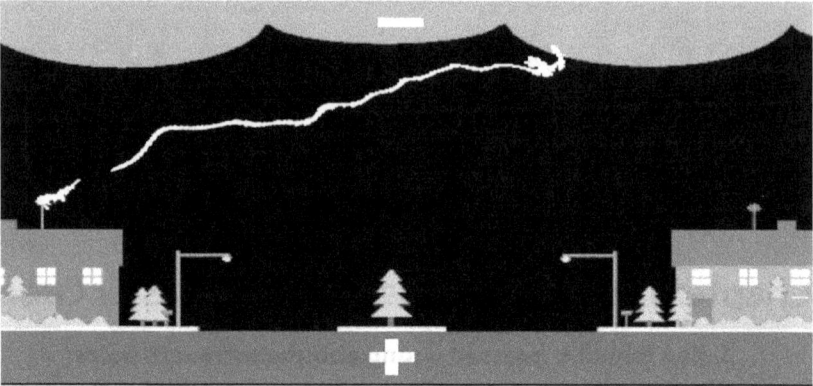

The mislabeling of electrical charge regarding its direction of flow

I suspect this habitual, sanctioned doublespeak about electricity's positive and negative polarity has caused some unconscious dissonance that still clouds science and causes it to misconstrue and confuse some aspects of polarized charge in general...for instance, in photons, neutrinos, gravitation, and dark matter. However, at the mobic scale, the polarized charge that generates space and time is not electrical; it is gravitational, and we'll try not to mislabel or misinterpret its polarity, which is more complex than that of electrical charge.

2. The tetrastar

I cannot show you photos of 3D space and 3D time latticing in the mirror-twin bubbles, nor can I show you meter readings taken on the polarized bonds that create scaling in the dimensional lattice. But I can describe it symbolically. The universe's dimensional latticing is based on its first tetrahedron, and its pulsing, polarized bonds can be described by I Ching math. Now let's describe the lattice structure and deconstruct its polarized bonds.

In Chapter 7, you saw how the polarized tension path running around both sides of a 2DD triangle proposed the 4 bigrams. In Chapter 9, you saw how the first 3DD tetrahedron confirmed those 4 bigrams by consolidating every polarized charge on its four 2DD planes into the 4 bigrams sitting at its 4 points.

Why is dimensional latticing based on a tetrahedron? It's because—fortunately—unlike all the other Platonic solids, a tetrahedron is self-dual. In other words, the tetrahedron can have a reversing-mirror twin shape.

Lada Prkic, Head of Technical Development Department at the University of Split explains this idea on the website *Geometry, All Around Us*, "Each Platonic solid has a dual Platonic solid. If a midpoint (centre) of each face in the platonic solid is joined to the midpoint of each adjacent face, another platonic solid is created within the first.... The tetrahedron is *self-dual* (its dual is another tetrahedron), the only one with 4 faces and 4 points."

This *self-dual* tetrahedron also goes by various other names. Myself, I prefer to call it a *tetrahedral star* since all four spurs of one tetrahedron project from the four faces of its dual tetrahedron…and often I elide it into just *tetrastar*.

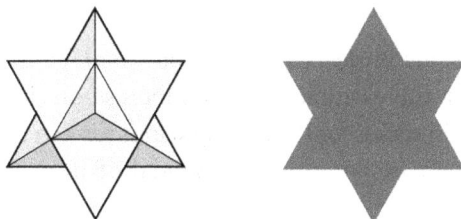

A 3D tetrastar has 8 points; a 2D Koch snowflake has 6 points

The 3D version has 8 points. (One is hidden in this image). And if you turn the 3D tetrastar so that a spur points directly at you, its silhouette becomes the 2D Koch snowflake. It has 6 points, and its fractal geometry was described mathematically by Swedish Prof. Helge von Koch in 1904.

To iterate the fractal form of the 2D Koch snowflake, start with an equilateral triangle. On each side, at its middle third, erect a smaller equilateral triangle.

Growing a 2D Koch snowflake

The graphic above takes this iteration through just four stages, but you could continue the iteration process onward. A 3D version would turn it into a spiky ball looking much like a medieval mace. Theoretically, you could keep iterating this fractal 3D shape to grow ever-tinier points until it is a nubby burr of points that vanish into mathematical infinity…theoretically.

3. The tetrastar in a cube = tetracube

You already saw how tetrahedrons project their two volumes—one of 3D space, the other of 3D time—outward beyond the mobic scale to form hourglass cells merging into the holographic bubbles of 3D space and 3D time.

Next, to establish dimensional latticing on out beyond the mobic scale in the 3D space bubble and 3D time bubble, each bubble developed *self-dual* tetrahedrons whose edges sketched cubes in an invisible structure that grids the lattice and holds it together via flex-bonds that hexagrams can describe.

How do the tetrastars operate in the lattice? When a tetrastar fits into a cube, it becomes a *tetracube*. All 8 star points exactly meet all 8 corners of the cube. Each star-point bonds three poles of polarized tension. So does each cube corner. Those star points and cube corners polarize on the same exact match. Their shared bonding points can be expressed as trigrams on the 8 points/corners.

On this tetracube image, Point #1 is obscured. Point #7 is difficult to differentiate. So in this example, those numbers sit on circles to make them more evident. If yin stands for 0 and yang stands for 1 in binary format, all 8 trigrams count out in binary sequence around the 8 tetracube points-corners.

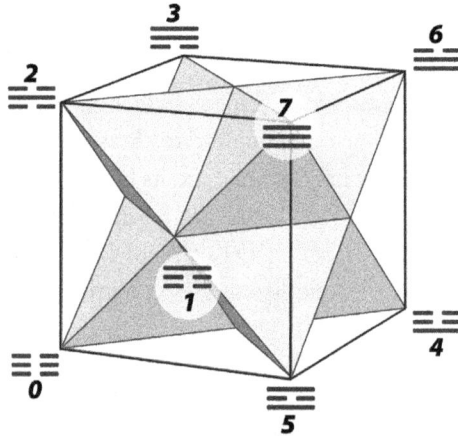

Tetracube: 8 star points, 8 cube corners, 8 trigrams

Now let's deconstruct this tetracube to examine its polarity. For clarity, I'll make its tetrastar interior invisible, but of course, it's really still in there, and remarkably, all of the trigram polarities we'll explore apply to that tetrastar, too.

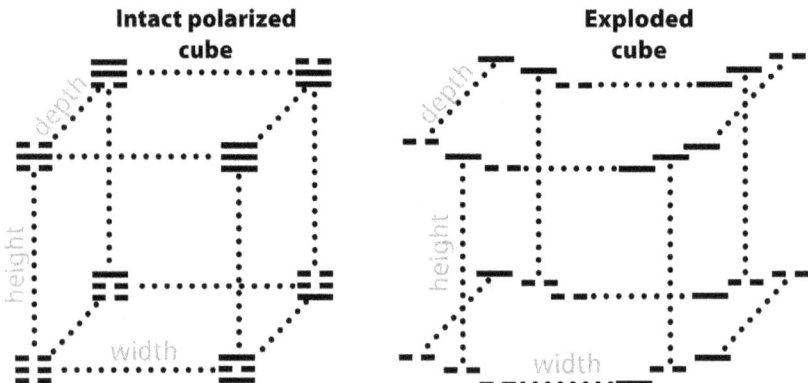

Intact polarized cube

Exploded cube

The 8 corners bond into the 8 trigrams...

...that mesh the 24 poles of cube tension

Deconstructing a cube of dimensionality and sorting out its 24 poles

Above left, you see an intact cube with 8 corners. The cube's 12 polarized line segments are 12 polarized tension paths, expressed here as dotted lines. Each dotted segment has 2 gravitational poles. Above right, you see the deconstructed cube. It shows how to sort and bond all 24 poles into 8 trigrams.

To establish a trigram at a corner, 3 poles on three different segments must bond. Since all 8 corners have 24 poles, they bond by triplets to put 8 polarized trigrams on the 8 corners of the cube (and star points).

A simple algorithm sorts each pole into its own trigram:

Width pole = *bottom line of a trigram*
Height pole = *middle line of a trigram*
Depth pole = *top line of a trigram*

When the polarized cube is put together, how do its parts relate in terms of polarized strength? Below right, you can follow the binary counting sequence from 0 through 7 around the cube to find that its path traces two big, complementary **Z** shapes on the left and right sides of the cube.

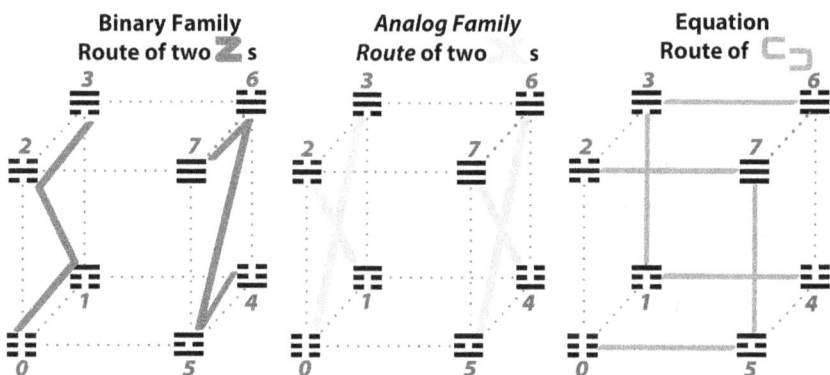

Alternate routes around the tetracube can test its polarized strength

Shao Yong's binary trigram order traces two Zs on the cube

The middle cube shows two big **X**s on its left and right sides. They track King Wen's analog order of trigrams seen below.

King Wen's analog trigram order traces two Xs on the cube

You can also make equations of various polarized symmetries, such as…

Trigrams can also trace various polarized symmetry equations on the cube

4. The 8 trigrams = 8 vp3s of polarized gravitational force

Earlier chapters in this book showed that simple yin and yang polarity develop at the first forking of a p-tree. At its second forking, bigrams develop. Just one more level of polarized bifurcation generates all 8 trigrams on the p-tree. Reading from left to right, they count in binary from 0 through 7.

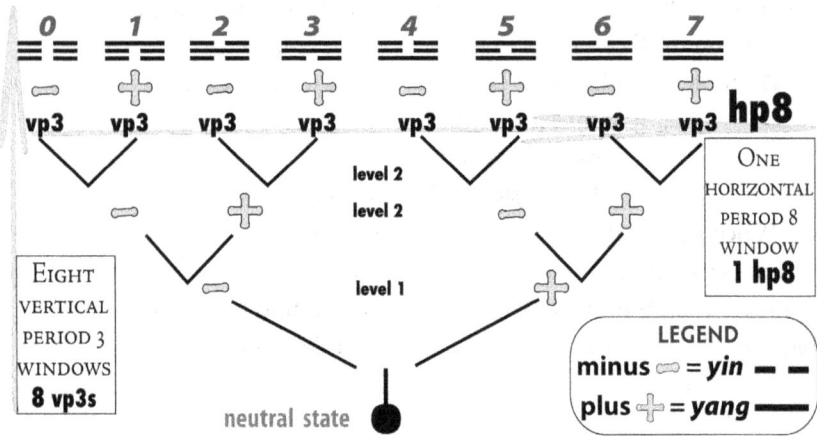

Trigrams appear at the 3rd level of forking

However, mathematically speaking, there is much more than just binary counting going on here. In 1975, James Yorke and Tien-Yien Li's pivotal paper, "Period Three Implies Chaos" proved mathematically that when a *horizontal* period 3 window appears in a bifurcation-tree, or b-tree, above that level, the branching no longer records mere random chaos. Instead, above that level, the b-tree now records a dynamic of fractal, patterned chaos. It was a great find in chaos theory, and Book 1, Chapter 9 discussed some implications for this TOE.

A brief review of those implications follows. This TOE proposed that a polarized bifurcation tree, or *p-tree*, if read *vertically* up just three levels, delivers 8 vertical period 3 windows (vp3s) by the third level of branching. The p-tree's 8 polarized vp3s are symbolized by the 8 trigrams counting in binary order.

In a trigram, each pole (symbolized by a yin or yang line) generates its own wavelength or "note" of resonance. All three levels together become something like a chord in an octave of energetic possibility. All 8 trigrams describe all 8 possible vp3s. They deliver 8 different harmonics of gravitational chords.

The p-tree not only counts in binary numbers; it also juggles the analog ratios of polarized relationships inside each trigram and across the whole p-tree. Its branches hold a symphony of wave relationships that act very different from the linear sequencing of a binary path. Analog waves of polarity have a different purpose from binary chunks. Their gift of resonant relationship in the master code is what lets the universe's parts relate in groups that alter their dynamics, not just pile together in unrelated lumps. Mixing chunky units with relational waves allows the 8 trigrams grown on a p-tree to develop the special, specific kind of nonlinearity that I call analinear.

Each trigram is a fractal within its family of fractals. In summary, the p-tree not only produces the binary order of trigrams in a mere three levels of forking; it also provides 8 different vp3s of relational, analog resonance. What a compact mathematical notation is this I Ching shorthand!

5. Paths on the tetracube reveal mathematical properties

Now let's put the tetrastar back into its cube and place on its 8 corners the efficiently consolidated polarities called trigrams. Simple stages of increasingly polarized bifurcation are what developed this first fractal tetracube.

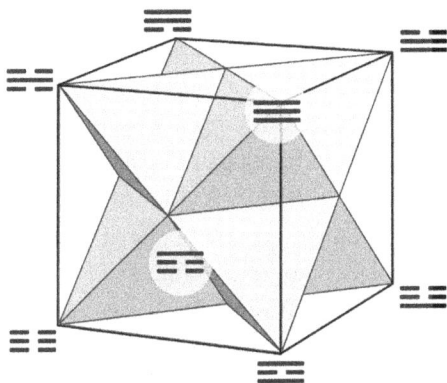

The paths of **2** reinforce each other

2-*Adding*	**2**-*Doubling*	**2**-*Exponentials*
2	2	$2^1 = 2$
4	4	$2^2 = 4$
⑥		
8	8	$2^3 = 8$

Trigrams appear at the 3rd level of forking

Studying the tetracube above, consider how the accord of its unique analinear math path from 2 to 8 allows the fine-tuned Double Bubble universe to exist. Book 2, Chapter 13 pointed out in detail how the co-chaos paradigm relies on the number path from 2 to 8. Why? Because the 2 can take a chunky, unitized path to reach 8 by **adding** itself ($2+2+2+2 = 8$)...or it can take two different analog, relational paths to reach 8 by **doubling** itself (2 doubled = 4; 4 doubled = 8)...or by doing **exponential growth** ($2^2 = 4$; $2^3 = 8$).

Its additive route ($2+2+2+2 = 8$) must, however, make an extra stop at 6, its hiccup spot circled on the chart above. Still, all three methods reliably hit 2 and 4 and reach the goal of 8. No other number path has three alternate routes that boast such a tight and total accord of method, progression, and goal.

The 2's ability to take three different, reinforcing paths to reach 8 establishes a mathematical merger that is nearly magical. Each trigram describes a unique fractal chaos pattern whose number surety is secured by its three internal paths of 2 moving to 4 and onward to 8; so many options reinforce each other's accuracy. This combination of binary counting/addition/doubling/exponential growth creates a tight, nonlinear result so special that I call it analinear. Its merger of method, progression, and goal between 2 and 8 is so versatile, so widely applicable that, for example, this same number surety used in the genetic code lets DNA remain stable as it meanwhile paradoxically also evolves.

This number merger is also what makes the star-in-a-box geometry so magically sturdy. For instance, we know that in our upper bubble's 3D space, a tetrahedron's volume is precisely ⅓ of the volume of the cube around it. Yet strangely, a tetrastar made of two tetrahedrons fills only ½ of the cube's volume, not ⅔ of it. That's because parts of both tetrahedrons interpenetrate.

So you could actually pour the water volume of *two* tetrastars into this cube that holds only one tetrastar. In other words, the volume of two tetrastars formed by four tetrahedrons equals the volume of three tetrahedrons. Huh?

All of this seemingly odd chance holds a hidden fractal promise. It packs the numerical clout of a 4-ness into a 3-ness born of a 2-ness. Huh? It also verifies the Rule of 4's insistence that when a threesome appears, its threesome will drive toward finding an ornery fourth to initiate a pair of pairs. Huh?

And it's all due to a 2DD triangle finding that original fourth point to develop a 3DD tetrahedron. The insistent fourth point brought an essential difference that pushed the system to evolve. Like John Lennon did for the Beatles. Or like female Mary did for the male Trinity. That's also why for stability's sake, rather than a threesome, it's wiser to double-date or to golf in a foursome. All of this honors an archetypal dynamic born from polarized pulsing on the path of a closed orbit around either side of the 2DD triangle. As Maria Prophetissa said, "Out of the 3rd comes the 1 as the 4th."

The tetracube holds the tetrastar. The trigrams bonding its combo-structure describe its polarized relationships. They explain the mathematical surety of structural integrity existing in the Double Bubble's dimensional latticing. This single polarized tetracube, when multiplied into many, becomes the holographic, "squashy" latticing of tetracubes that allows the scaling ratios of change to operate in the Double Bubble's dimensionality.

The polarized trigrams on tetracubes bond the dimensional latticing of both bubbles into a complexly polarized system that stays strong and enduring via its invisible bonds. It can maintain structural integrity despite the lattice's successive iterations of scaling in ever-larger or tinier tetracubes.

6. Hexagrams bond the dimensional lattice

The Double Bubble's space-time lattice is structured according to inherent fractal traits. In this upper bubble, those traits let our 3D space and ½D time display diverse objects of 3D mass glued together by energy—a molecule, a mountain, a man. The mass is ever-emergent in spatial reality, ever-renewing, yet also changing along ½D time's omnipresent, moving point of *now*. Meanwhile, the lower bubble lets its 3D time and ½D space cooperate to evolve the unified mind's continually emergent 3D patterns of tachyonic energy moving on its omnipresent point of *here* in ½D space. (There is no *there* there.)

Earlier volumes in this series (especially Volumes 1 and 2) examined the p-tree growth of co-chaos, shorthanded by I Ching math. Here's a brief reprise:

The p-tree can sprout roots as well as limbs to become a double p-tree, or *dp-tree*. Its three levels of branches and roots, up and down, grow 2 sets of polarized 3-packs. Each 3-pack is a *trigram* with its own unique chaos dynamic. All 8 × 8 trigram 3-packs can pair-bond across domains into 64 *hexagram* 6-packs that describe complementary chaos…or for short, *co-chaos*. Each hexagram 6-pack describes its own unique co-chaos dynamic.

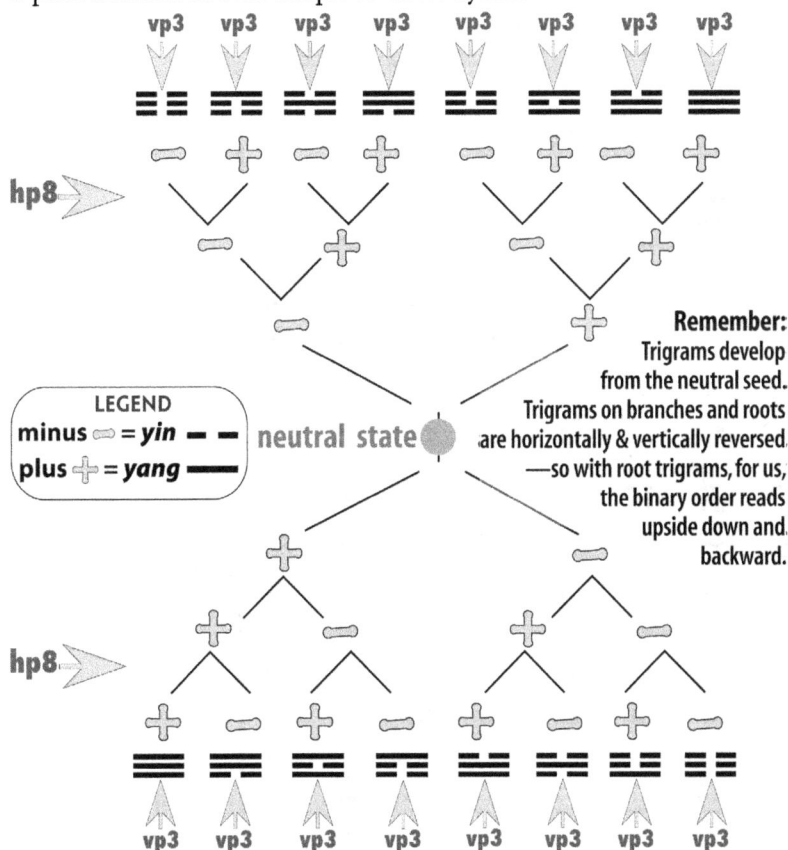

LEGEND
minus ⌾ = *yin* ▬ ▬
plus ✛ = *yang* ▬▬▬

neutral state ●

Remember:
Trigrams develop from the neutral seed. Trigrams on branches and roots are horizontally & vertically reversed —so with root trigrams, for us, the binary order reads upside down and backward.

This dp-tree has 8 × 8 vertical period 3 windows (vp3s)

By growing just three levels up and down from its 0-seed, the dp-tree's polarized complexity has moved far beyond what an ordinary b-tree can muster in standard chaos patterning. A b-tree forks into a multitude of twigs that develop just one hp3 on a given level. By contrast, by just the third level outward, up and down, a dp-tree grows 8 vp3s. Each vp3 packs more punch than a b-tree's hp3 can muster, for the vp3 supports both binary data and analog resonances within the co-chaos patterning of its dynamical system.

The merging of binary data and polarized resonance in each hexagram offers a big bonus. By combining both functions in its shorthand, the I Ching math can symbolize the merger of binary chunks with waves of resonance to generate the 64 analinear patterns of varying energetic consonance and dissonance, along with an inbuilt binary count.

The result? Pair-bonding all the possible trigram combinations delivers 64 polarized 6-packs of co-chaos dynamics that hold together the Double Bubble universe's dimensional latticing.

I Ching's 64 Hexagrams in Binary Order	RNA's 64 Codons
0 in binary code	UUU UCU UAU UGU
	UUC UCC UAC UGC
	UUA UCA **UAA** **UGA**
	UUG UCG **UAG** UGG
	CUU CCU CAU CGU
	CUC CCC CAC CGC
	CUA CCA CAA CGA
	CUG CCG CAG CGG
	AUU ACU AAU AGU
	AUC ACC AAC AGC
	AUA ACA AAA AGA
	AUG ACG AAG AGG
	GUU GCU GAU GGU
	GUC GCC GAC GGC
	GUA GCA GAA GGA
63 in binary code	GUG GCG GAG GGG
	U = Uracil A = Adenine
	C = Cytosine G = Guanine

The 64 hexagrams sitting in parallel with RNA's 64 codons

In Book 3, *Tao of Chaos,* I mentioned that you'd see again this chart of binary hexagrams cross-correlated with RNA codons. It parallels the I Ching's 64 hexagrams of ancient China with the 64 RNA codons of modern genetics. On the left, the binary order of 64 hexagrams uses yin to stand for 0 and yang for 1 as it counts in binary from 0 at the upper left corner of the chart across and down to 63 at the bottom right corner. On the right sit the 64 RNA codons. Their amino acids make the building blocks of all organisms.

The I Ching's math and text have been studied for at least three millennia. The genetic code was decoded only in the 20th century. We learned how it sets information into DNA and then decodes it into RNA's amino acids that make proteins in living cells. Book 3 showed you how both systems, genetic code and I Ching, are based on the same underlying paradigm.

Seeming differences in the charts merely camouflage their cross-correlation at a deeper level. Yes, the I Ching chart has 8 columns × 8 rows of yin/yang lines, while the RNA chart has 4 columns × 16 rows of alphabet. And yes, the I Ching uses only 2 linear symbols (yin and yang), but RNA uses 4 molecular symbols U, C, A, and G (substituting Uracil for the Thymine found in DNA.) And yes, a hexagram has two 3-packs that bond into a 6-pack of lines, but an RNA codon is just a 3-pack of molecules.

Nevertheless, Book 3 showed how those two systems are two different ways of symbolizing the same underlying dp-tree of polarized math in a co-chaos system. Examining both systems carefully invites a hypothesis that the genetic code and the I Ching math are actually two fractal variants of the same underlying co-chaos paradigm.

This Double Bubble TOE proposes that the genetic code and the I Ching math are two evident but lesser variants of a deeper master code that set the paradigm. It suggests the master code developed dimensionality emerging at the mobic scale in the Double Bubble universe and loaded it with particle-waves emerging at the quantum scale of this upper bubble. Scalar change occurs in its polarized dimensional latticing, and hexagrams can code it.

This TOE says we are presented with a Rosetta stone of three codes, two known and one unknown. The two known codes are the I Ching and the genetic code. This TOE says the I Ching offers a mathematical key to open both other codes, the known genetic code and the unknown master code that is still obscurely hidden in various aspects of physics, math, and metaphysics. This TOE says that I Ching math can help us spot and decode the basic dynamics of the master code generating this universe.

This TOE proposes that the fractal master developed polarized pulsing and iterated it to build the space and time of both bubbles, leaving in either bubble a polarized debris of matter or antimatter as construction trash. Fortunately, all that litter in our bubble got repurposed into galaxies, suns, planets, and even us diverse little organisms with our small, walkabout minds.

More fortunately for us, the antimatter debris in that other bubble got crush-converted and repurposed into speedy tachyonic energy that powers a single huge mind. That unified mind in the other bubble is what we call Mother Nature. It slowly made this bubble liveable by working over time through the mobic pores of the membrane interface, rather like building a ship in a bottle.

And even more fortunately, this living universe still keeps refreshing its durable fractal carriers—space and time—and updating their variable contents—matter and energy—in the continually emergent solution that we experience as events.

The analinear dynamics of co-chaos let the master code coordinate polarized information at the mobic scale to create and maintain the space-time lattice, plus schedule matter and energy to morph up here above the quantum level in our own upper bubble of this holographic Double Bubble universe. The lower bubble meanwhile manifests tachyonic energy with imaginary-number mass.

Imagine! There's a symphony of polarized pulsing at the mobic scale that generates the universe and updates it. And the I Ching math can shorthand it! By viewing nothing as something—the number 0—the cosmegg managed to develop dimensionality and mattergy to project a universal hologram of tiny events inside huge events stretching to the far horizon of what we call reality.

And there's a lot more we cannot see with our technological tools. The only way to probe the lower bubble is with the mind's eye using math, meditation, dreams, deep-see diving, out-of-body experiences, Akashic records, astral travel, altered states…whatever manages to take you into a place where you can experience what elephants and giant octopi with their less elaborately defensive egos already source: the mind of our universe that loves its creation.

In this mighty opus, the space-time skeleton and its matter-energy flesh form our hologram that is real by every definition we know. Its analinear dynamics allot a measure of free will to each bit of organic life and even give some leeway for mistakes, for how else could God as Grand Organizing Design find out what really works best as the instigator of evolving consciousness about love in this universal petri dish?

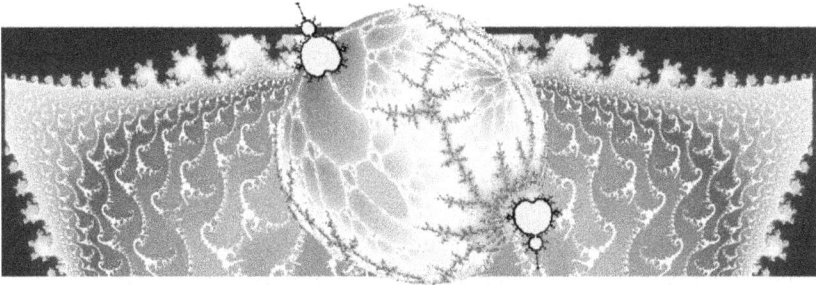

Chapter 14: The Good-Enough God

1. Our universe is a go-getter

This universe still lives because it chose to live, because it worked hard to achieve sustainability, and because it has plans. It is continually emergent, constantly renewing itself, updating its matter and energy contents within its dependable space and time carriers.

Most of the universes that are still extant (and there are many) do not actually live, much less evolve with a concerted focus on consciousness. They just click along mechanically, and it usually is not by means of time. However, the Double Bubble does not grow stale or stiffen up as so many other universes have done in the original spawn, expiring or already dead around our own universe. Instead of being automated, insensate matter, our own living universe was generated by an analinear master code that allowed it to stay flexible enough to adapt, evolve, and even explore consciousness.

As the maturing universal mind in the lower bubble increased in complexity, it tinkered our upper bubble's plasmic gas into stars and planets, then coaxed forth little organisms such as us evolving along the arrow of time. The DNA spiraling in each cell of your body helps evolve the species through successive generations, much as each iteration of polarized pulses at the mobic scale helps evolve the universal body.

Polarity's connective resonance shows up even in the cultural patterns that bond us at a visceral, unspoken level, and they mesmerize us. They offer the kinetic impact of call-and-response, of palm-on-palm contact, of many bodies moving with a shared purpose. The polarized *do-si-do* in a square dance echoes that yin-yang polarity, and the four bigrams spell out the dance's pairing options in the simplest possible terms.

2. Love has the greatest survival value

Why did the cosmegg begin to make something out of nothing? Why did it bother to elaborate the four primals into a holographic Double Bubble of space and time with a constantly varying contents of matter and energy?

Why did the universe move toward greater complexity and awareness among its parts, coaxing atoms, molecules, stars, planets, and organisms to develop? Why, working from the lower bubble, did that vast mind sculpt in this upper bubble the myriad tiny organisms carrying portable mini-minds? Why did it pack life so tightly that your own body contains 10 times more microbial cells than human cells?

Why did that great mind send our limited minds various ideas "out of the blue" to develop habits and tools that provoke more awareness of ourselves, even to speculate on something greater than ourselves? Perhaps even to develop a degree of awareness of the universal mind residing in the universal body?

Upon meeting it, some call it Mother Nature. Others suppose such wisdom must be God. However, the universal mind is merely God's minion, an aspect of God, just as we are. Why does that great mind still keep rearranging things? Because it seeks to develop its own potential by relating to the diverse little minds inside it...like us...and also to that which is greater beyond it...like God.

Okay, if that is the plan, then why not coax those tiny, perambulating minds in the upper bubble into training camps, so they can learn to live for something more than just themselves? How?...let's see...how about letting them acquire mates? Families? Tribes? Societies? Nations? Religions? Dreams? Sciences? Extraterrestrial trips? Intergalactic instant messaging among distant beings? Deep-see diving that can lead to accounts such as this one?

Meanwhile, in the tachyonic cloud of the lower bubble, the symphony of resonant activity in 3D time keeps increasing as the huge, unified mind explores events that tweak more meaning into life. Our remarkable universe cultivates awareness within its massive body—sensory, intellectual, intuitive, emotional, spiritual, and other kinds of awareness that I cannot tune into or name because that bandwidth is so much broader than mine.

But I do get bits of it cutting in and out. I can tell you that it nurtures us through the selective cultivation of that which is functionally able to survive long enough to get wiser about love. Not tougher, smarter, meaner. Wiser.

Oh, the universe loves this plan! It will work. It will give those tiny mites of consciousness some training wheels. Let them discover every shade and resonance of love and its opposite, how to reframe that music again and again into something that is truly viable, negentropic, effective over the long haul.

Aha! It will give love the greatest survival value of all. Consciousness in our particular universe will live for love, die for love. It will assay every kind of love—tough love, romantic love, resigned love, sexual love, greedy love. It will plumb egomaniacal fixation and selfless devotion. It will explore the love of truth, of sadomasochism, of addictive substances, of integrity. It will test every shade and resonance of love and its seeming polar opposite, hate, evolving love itself by finding that which works best, again and again over time and space…until all the vibrating overtones of love become reconciled enough to admit that even hate is love gone wrong…until loving life and embracing it well, truly well, is refined, purified, distilled, alchemically transformed into the best essence that this universe can offer up as that which loves its originator, the Grand Organizing Design, which I will for convenience call God.

3. Time puts us on the straight and narrow arrow

For us here in the upper bubble, time as a teaching arrow to evolve us in 3D space has worked out well. We've even learned to stretch time's limits in our lives somewhat by using airplanes, computers, media, and medical discoveries to crunch more events, facts, opinions, feelings into our days.

But we still cannot jump into yesterday to derail an event that has already happened so that it never even occurs. The relentlessness of this ½D, one-way timeline is what keeps our tiny minds from getting too cocky. At our current stage of awareness, it would ruin the Tiny Minds Learning School to allow the unjust among us to wreak havoc in 3D time as some already do in 3D space. Too many of us still treat life as if it were a game designed to trick, cheat, defeat, and kill each other in ego battles.

If we all were given blanket access to move around in 3D time as we already do in 3D space, it would skew the final outcome too much to make that risk worthwhile for our universe, so intent on consciousness as the key to refining love. Many would become Machiavellis of 3D time, given the chance, caught up in manipulative schemes, turning that current life into a warped, broken bead on the soul's ongoing necklace of many lifetimes.

Which is not to say that we humans will always remain so time-limited. If things go well, our whole species will evolve by gradually expanding our range of vision in time. As a species, we'll exist more consciously aware of time as a teacher until we rejoice in the learning life of our universe itself, loving it onward toward its next great stage of being.

4. Living with the good-enough God

Why do we need to acquire so much wisdom about love before we can get much closer to God? Because we need it to accept God better. Unlike what we fondly wish, I am told, God has not made this universe perfect. Nor is any universe perfect. Each has some limitations automatically built into the parameters of its construction. That is what makes all universes finally have an implicit desire for completion in God. They seek wholeness in that which is God.

God is creative source itself. Not really a he or she or it…those are just limits of language and thought. God is all and wants to know us in all God's parts and chambers. But exactly how to do that is like a big puzzle that keeps changing its access code with each universe. Why? Because everything that the Grand Design builds must be constructed in something, be it the matter we know or something much more shimmery and evanescent...or less.

In any case, existence must be meted out in some fashion in each universe, with rules of standardization or givens of some sort...and with parameters come limits for whatever is created in a universe and how things interact within it. For instance, you know how hard we humans are to please. Yet also how hard we aim to please.

Moreover, even God does not wish to be taken for perfect. God has lacks and sorrows of a kind that we do not know, I am told, but they are real, and God wishes to be considered merely good enough, an awareness that is trying to test love as the greatest value in this universe. As we realize how difficult it is to love really well and manifest it really well, refining love is our task. Thus in a strange way, we are refining God. Just as we do better or worse with our lives, so will God do better or worse with creation, but on a far grander scale that is incomprehensible to us. God is refining us, and by it, we are refining God.

Instead of alternately claiming that God is perfect or blaming God, simply know that this good-enough God does what is possible for us each day, given the parameters of our universal setup, our personal circumstances, and our free will. And know that some of the burden rests on us, too, for maintaining a good-enough situation that works long-term for its inhabitants.

We can take responsibility for our own small part in the design and do our best by it, using regard for ourselves, for each other, and for the wonder of our source. God doesn't make things instantly right and ready, shaping it up to be comfortable, kosher, halal, copacetic, A-OK, and peachy-keen without our help.

True, God's fingers snapped and sparks appeared—well, really, it was parts of Godself that came flying out to jump-start this spawn of universes. Then they all began sorting out according to the predisposition in their separate seeds. In this Double Bubble universe based on a seed of "nothing" that saw itself as the number 0, with the possibility of binary and bifurcating, polarized numbers, we must experience life sometimes in a negative aspect.

Sometimes we even experience the divine that way. We call it Satan or the Devil, or bad things happening to good people, or like a furious Christopher Hitchens, insist that *God is not Great*. But God is great. Big and wondrous, huge and divine. God is something far larger and grander than any universe, a creative source that emits such a grand organizing design that it can even meet all of us in different ways personally as we are ready—whether we expect it or not, want it or not, recognize it or not, admit it or not.

In a weird way, God could be seen as the greatest narcissist of all time and beyond time, because everything reflects back to God, is a part of God. He's what it's all about. Let's talk about him. But of course, God is not really a him. The conventions built into the English language, and even into the Western mindset, cause God still to be seen as a Heavenly Father, a Lord over a medieval, invisible fiefdom, or a King with a gray beard and non-detachable throne.

But the God I know is Grand Organizing Design, a creative force so big that it generates everything to explore and refine new aspects of itself, even dressing up in flesh occasionally to participate in this design firsthand, putting us tiny mites into the loop of being as an aspect of its ongoing refinement, using aspects of itself to talk to each other and to God, because the Master Designer is continually reflecting upon the design to improve it in truly viable ways.

That is what the whole universe aims for…reflection on aspects of the source that refine the source. It fosters multiple images for reflection in our hologram—physically, mentally, emotionally, culturally, spiritually. It mirrors one aspect upon another in polarized ways. By reflecting on those polarized, multiple views, consciousness arises in many forms that can begin to sense the shimmering presence of a higher power. We come to divine it. It aims finally toward showing divinity bare, whole, and remarkable in all its parts.

All of the parameters in our own universe help God know more about the utility and limits of love. By becoming so deeply aware of the limits of love and the havoc that it wreaks in this universe, we are a godsend. By watching our antics, God is trying to spot inadequacies regarding the manifestation of love in this universe and work on it right here, right now, with us, since without any peers, it is hard for God to get any mirroring done otherwise.

Although we are not God's peers, by loving us despite our flaws and despite the limits that are built right into this manifesting universe, God is showing us how to love better. We are doing the same by treating each other better, and by that, we even love God better and so improve upon God's love quotient.

5. Riding the lightbody vehicle in the good-enough God

Many here in the upper bubble already know, consciously or unconsciously, of the tetrastar dynamic that exists at the fundament of nature. We recognize it at some intuitive level beyond logic. It is part of the perennial philosophy. We have given this tetrastar many names in various cultures across recorded history. Humans worldwide have drawn pictures and performed rituals based on the shape of the tetrastar, using variations of that image to honor a projective, holographic dynamic created by polarized beats at the mobic scale.

Tetrastar names: MerKaBa, star tetrahedron, stella octangula, duo-tet

In ancient Egypt, it was called *MerKaBa*, three words in one that meant the Light Body Vehicle. *Mer* is a rotating field of light; *Ka* is spirit; *Ba* is the body. In Hebrew, it was called the *Merkavah*. A 2D version of it was called the Star of David. In mysticism, it was known as the hypnotic *star tetrahedron*. In 1509, mathematician Luca Pacioli put a *stella octangula* in *Divina Proportione*. Over 400 years later, Buckminster Fuller in *Synergetics* called it the *duo-tet*.

How long will the dimensionality in our Double Bubble universe be generating these tetrastars that project the holographic lightbody vehicle in which we ride? That ride will not be endless, but it will not end in a boom or a bust, nor in the uniformity of a cold dark soup, nor in a hot and heavy crunch. It will not become a forever-expanding void or a hellish knot of nevermind.

This ride will end as it arrives somewhere else. Our universe will cease when... if...as...it cocoons up like a dense foam bolster and evolves into something else, morphing into a huge, filmy light body that is evanescent, yet nearly eternal. Something much more fun and light-hearted, cashing in the bonus on all those light-years of effort in exploring love for the Divine Presence beyond all.

Our universe is heading for reincarnation as a light body for spirit embracing both 3D space and 3D time together. That meshing of dimensionality will allow its souls such an evolved version of being, give them such a high, wide, and handsome range of exploration in a vast new array of options that it hopes to grow God's being in new ways that will intrigue, inspire, and delight all.

6. Tips for encountering the divine

If you wish to encounter more divine presence in your life, I suggest something simple. Look inward and refine your character. I cannot tell you precisely how to do that. Your society, your associates, your own ethics—such pressures have unique ways of testing your character. Try to score well on those tests.

Meanwhile, observe your own emotions. Discover how they affect your behavior and perception. Coax them toward promoting good intuitive leaps, not poor ones. Watch to see how well you understand your friends…harder yet, your family. You are teaching yourself to notice subliminal cues that are embedded in the world around you, including those beyond the normal senses.

Stretch your intuition to sense things beyond your normal range of knowing. Keep testing your surmises against the emerging reality. Educate your intuition in simple ways to find out if you can, unbiased, predict 20 minutes into a movie how the plot will go, or how your tax audit will turn out, or which book your child wants you to read aloud at night. In an unfamiliar restaurant, guess if the restroom you are seeking will be on the left or right.

Notice your accuracy rate and try to improve it. Always test each hypothesis to see how it actually plays out in real time. If you recognize where you went wrong, acknowledge it. Immediately. That's the only way to shift the feedback loop into greater accord with the way of the Tao.

Spend enough time in nature to let you start feeling and adjusting to its deeper wisdom. Try several styles of meditation to find out which one works best for you. Then practice it until you reach a satisfying new place in yourself, in your life, until you inhabit a better state than you even knew existed before.

Pay attention to each dream, no matter how fragmentary or silly-seeming. Ask why it came to you just now, what it signifies in your life. To untangle a dream's details, look below overt events to discover the underlying symbolism. Deconstruct its *who, what, when, where, how,* and *why* to recognize that its meanings go deeper than cookbook-style definitions may suggest. For instance, black cats are not necessarily bad luck, and your nightmare of death by fire almost never means that you will literally die in a fire, but rather, that your ego identity, the "I" walking around in the dream, is undergoing a trial by fire.

You can figure out a lot just by asking the universe to trigger synchronicities in your ongoing life to inform you, transform you, help you on a need-to-know basis. Then stay open to the cues humming in events. The watercourse way inevitably seeks the easiest, most natural path for your life to reach its true goal… which may be rather far from what your ego deems ideal or even desirable.

Back when I was an indifferent agnostic, I did not consciously choose to consummate a dreamy union with God. I did not ask for a nightsea voyage going back to universal creation and beyond. I was merely trying to understand myself better by going to weekly sessions with a Jungian analyst. But then that amazing God dream came along and sent me searching for its meaning. It led me to change my path and eventually write this series. It entrained me in a dynamic I could not resist, did not want to resist, even though it redirected my life and at times led me into more pain than I am willing to write about here.

Life takes us to strange places beyond our conscious choosing. Change what you can for the better, and make the best of what you cannot change. For instance, one does not choose a fatal illness, but many people report appreciating life the most, savoring its gifts the best, just as they are on the verge of leaving it.

Most important, refine your heart. I do not mean you need to become a Goody Two-Shoes, a soft touch for the con artist, or a weepy wimp. Just open yourself to love in ways that prove to be healthy and happy for you and those around you. Tap into the possibility that something beyond your logic can guide you wisely; something can see around the corner of time to discern how an apparent loss—of a job or a close friend, say—also opens an unexpected door.

Reframe your perspective to make each day become a precious gift to be treasured and enjoyed. It is not just a matter of finding the silver lining in a dark cloud. Rather, it is realizing that a dark cloud can turn golden.

Medieval alchemists had a motto: "Turn shit into gold!" They meant it literally. They tried to use the hot, fertile energy of decaying manure to morph lead into gold. Carl Jung realized those medieval alchemists invested so much thoughtful effort into transforming lead into gold that they began over time to transform themselves. Due to the hopes and fears they projected into their work, it became not a process for making gold but instead for refining themselves.

Jung said life often wants us to recognize and acknowledge the gray, leaden truth of some nearly unconscious, ulterior, shitty motive submerged in ourselves. We can notice our own psychology enough to start to change it.

Today Jungian psychologists use the alchemical tag, "Turn that shit into gold!" as a directive to take the shitty events of your life and deal with them in a way that transmutes their value into pure gold for your growth. Events that at first seem horrible, unbearable, in time can bring you wisdom beyond price.

And in fact, that alchemical shit did indeed turn into gold. Medieval alchemy gradually turned into chemistry. "Alchemical process" became a metaphor in Jungian psychology for the psyche's journey of personal individuation. You are the container of this alchemy at work in you. It uses shitty events in an attempt to transmute your leaden miseries and depressive days into the gold of wisdom, integrity, kindness, love, and hope. Let life work on you, for you.

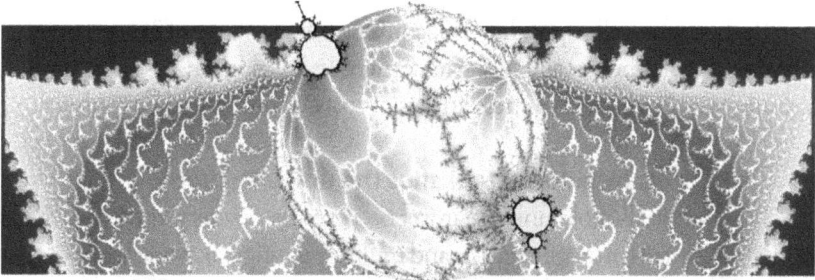

Chapter 15: The Good-Enough Universe
1. An overview of the good-enough universe

This chapter takes a last look through the scrapbook of our dimensional roller coaster ride. It's not a true summing up yet, just reminiscing about what we saw.

To think that it all started with "nothing" coming into being by realizing itself as a number: 0. Pulsing in and out of being, that pulsing beat becomes the binary 0-1 of *off-on* polarity. The 0 can also split and polarize into analog -1 and +1, another polarity. Those numbers develop a polarized pair of pairs.

At the mobic scale, the cosmegg grows into a 1DD line polarized by space and time; next, into a rapidly iterating, ultra-tiny triangle faced by 2D space and 2 D time; next, into a tetrahedron with an outer volume of 3D space and an inner volume of 3D time. The 3DD tetrahedron's quick-flashing 2DD triangles can remain at that scale, but its two volumes cannot, so it projects them as an hourglass cell whose wasp-waist sits at the mobic-scale middle.

AT THE MOBIC SCALE

This 3DD tetrahedron has four 2D space faces on the outside...

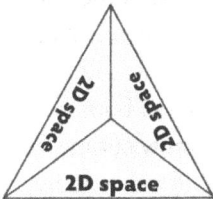

2D space
2D space
2D space

...& four 2D time faces on the inside.

CO-SIMPLEX GEOMETRY'S
SPACE-TIME DIMENSIONS

BEYOND THE MOBIC SCALE

It projects an outer volume of 3D space upward in scale...

Cosmegg

...& it projects an inner volume of 3D time downward in scale.

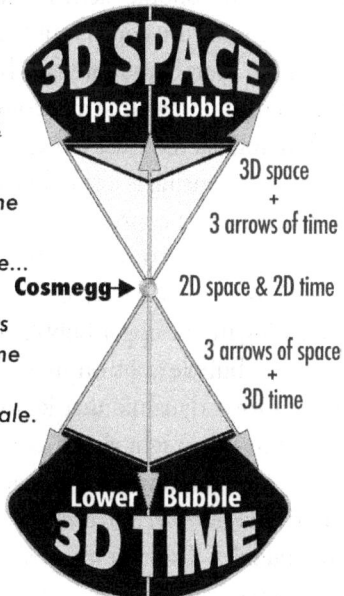

3D SPACE
Upper Bubble

3D space
+
3 arrows of time

2D space & 2D time

3 arrows of space
+
3D time

Lower Bubble
3D TIME

Two ways to symbolize a 3DD tetrahedron's 2 volumes

The hourglass cell's upper half holds an expanding volume of 3D space with 3 stress arrows of time running along its corners. Its lower half holds an expanding volume of 3D time with 3 stress arrows of space running along its corners.

2. Inflation establishes the universal body

The Double Bubble universe exists because the cosmegg used that single hourglass cell as a template to go big and multiply its original, workable hourglass cell in the "nothing" around it into many hourglass cells. In a flash, it abruptly became a huge organism with many cells! Physicists in our upper bubble call this huge growth spurt universal inflation. But this TOE says the inflation blew two huge holographic bubbles, one above and one below the mobic scale.

Suddenly the universe is now adult in size, if not in age or contents. Any residual gravitational pulses left over from the algorithm that established space and time got repurposed. Each bubble's space-time container received a load of contents automatically polarized to sort into that bubble. Our bubble above the quantum scale got original mattergy as a fine, uniform, plasmic gas. The lower bubble got original antimattergy, soon converted by the tremendous pressure of its ½D space into tachyonic particle-waves with imaginary-number mass.

The conjoined mirror-twin bubbles now exist as the white hole bubble of 3D space above the mobic scale and the black hole bubble of 3D time below it. Since they sit at opposite ends of the sizing spectrum for 3D space and 3D time, the two bubbles actually fit inside each other, producing a Klein bottle dynamic that turns our Double Bubble hologram into a kleiniverse.

In the membrane interface between both bubbles, each connective pore holds a mactor dynamic that constantly pulses along a 4-point orbital path on each triangular face just fast enough to perpetuate the 2DD triangles whose angular sequencing is just right to let lag time and lag space allow their fleeting bonds to form a constantly re-iterating 3DD tetrahedron, whose volumes of 3D space and 3D time constantly project above and below the mobic scale.

Meanwhile, a tensor network constantly 8-loops across both bubbles, presenting in our upper bubble as the omnipresent ½D arrow of time, but in the lower bubble as its omnipresent ½D arrow of space. If you consider only the arrowing dynamic that loops across both bubbles of the Double Bubble hologram, you might view it as a double torus. By taking a vertical, two-bubble cross-section of it, you'd find that each vectoring "arrow" in the double torus is part of an 8-loop in its tensor network moving on a continuous dynamic across both bubbles. The next image shows just two of the myriad 8-loops to suggest how the constant, omnipresent arrows create a tensor network that moves in only one direction—ahead—on a perpetual journey through both bubbles.

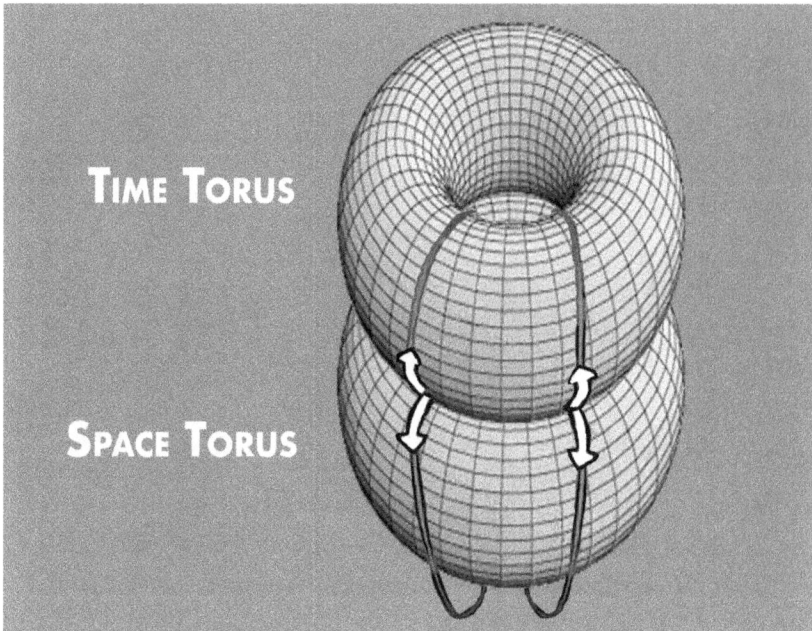

TIME TORUS

SPACE TORUS

The 8-looping tensor network of arrows creates a double torus dynamic

And whether the vector is coming upscale to the outer limit or heading back downscale to the lower limit of our bubble, it still reads to us as the omnipresent ½D arrow of time moving only ahead. Ditto for that arrow of space in the lower bubble. Thus the double-torus motion simply cannot go backward. Not now, not ever. That means our time can never go backward. (Too bad, sci-fi!)

Taken together, the pair of half-dimensions create a polarized, omnipresent 1DD line moving in an 8-loop across both bubbles. Everywhere. But…

space pole ⟷ **time pole**

The 1DD loop of tension

…the jazziest thing about this strange 1DD loop of tension vectoring through both bubbles on its endless journey is that it pays homage to the original, tiny 1DD line that our cosmegg first sketched to start dimensionality burgeoning at the mobic scale. Therefore, this looping paragraph now pays homage to that endless ∞-loop dynamic by imitating how….

3. Hexagrams can bond the dimensional lattice

Across both bubbles, numbers work together in analinear fashion to bond the elaborate scaling of dimensional tension in a lattice that crystallizes like an invisible skeleton in each bubble. I Ching math can depict its bonds.

THE CO-CHAOS DOUBLE P-TREE...   THE SPACE-TIME LATTICE...

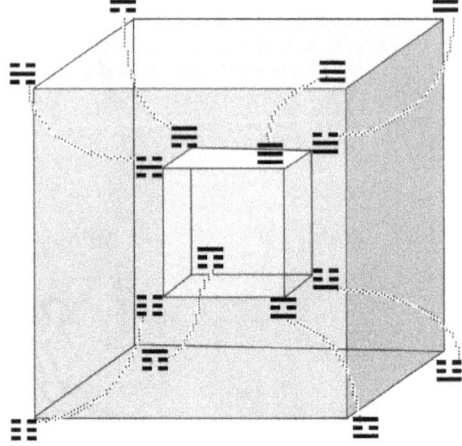

neutral state

⚌ = *yin*
╬ = *yang*

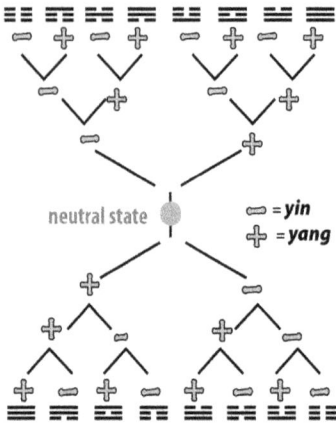

*...8 x 8 polarized vertical paths
equal the 64 hexagrams*

*...8 x 8 whirling tetracube corners
equal the 64 hexagrams*

The dp-tree grows the levels of gravity bonds in space-time's lattice

As two tetracubes of different sizes bond in the scaling lattice, their trigrams turn into hexagrams. Trigrams and hexagrams can chart the bonds of our 3D space lattice. Some of those bonds appear in the image below.

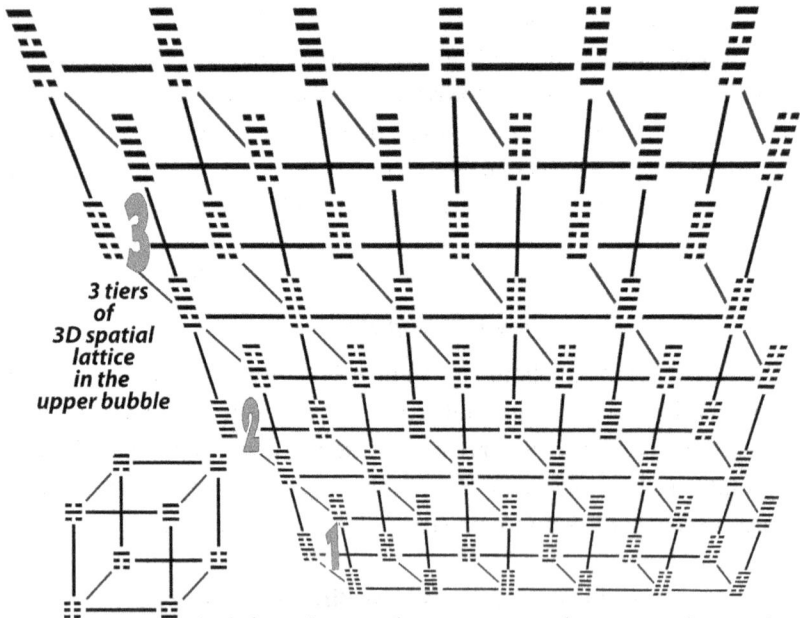

**3 tiers
of
3D spatial
lattice
in the
upper bubble**

**1 tetracube
of lattice**

And don't forget! The tetrastar girders are in these cubes, too, but just temporarily invisible for the sake of clarity.

A section of the dimensional lattice

This section of 3D space lattice in the upper bubble suggests how bonded tetracubes can adapt during scaling changes, using the bonds that hold the latticing of 3D space or 3D time together. Thus this theory is invariant under conformal transformations, i.e., transformations that preserve angles.

An elaborate dimensional lattice of flex-sizing establishes hexagram bonds extending across the Double Bubble. Scaling dynamics throughout the whole lattice allow the bonding polarities to shift and reintegrate with each change in scale and shift in contents. These lattice bonds are integral to the co-chaos patterns of mattergy on display throughout the Double Bubble hologram.

4. This TOE has some echoes of Nassim Haramein's work

It was only when I was starting to publish the 3rd edition of this series in 2011 that I discovered Nassim Haramein's work. His theory and this TOE seem to me like two different approaches to a holographic universe from somewhat related points of view. In Haramein's DVD *Crossing the Event Horizon: Rise to the Equation,* he describes a "holofractographic" universe with a mysterious "spacetime torque" based on the spin/angular momentum of matter and energy emerging at the quantum scale. But Haramein' describes just one torus field, relating only to matter and energy emerging at the quantum scale of what this TOE views as the upper bubble seen by current physics.

However, at least in some respects, this Double Bubble universe recalls a doubled version of Haramein's single torus of a "mysterious field in a tetrastar." This TOE describes the development of two holographic bubbles conjoined at the far-tinier mobic scale where space and time emerge. Gravitational pulsing in its membrane interface projects the two huge bubbles. Their two 3D volumes contain the double-torus momentum of ubiquitous, 8-looping vectors in the tensor network that we in this bubble experience as the arrow of time. This TOE, however, unlike Haramein's work, also declares that dimensionality, the genetic code, and I Ching are three fractal variants based on the same dp-tree paradigm, which bonds chaos dynamics by paired triplets into co-chaos dynamics that may be shorthanded by hexagrams.

5. Jamming the number lock on dimensionality

How many dimensions are in the Double Bubble? Count 3.5 dimensions per bubble, or 7 dimensions in both bubbles. Then add 4 more dimensions at the mobic scale itself, where each pore holds flashing 2DD triangles made of polarized planes of 2D space and 2D time…that constantly iterate another 3DD tetrahedron…whose four 2DD triangles can stay in the mobic scale, but whose two volumes cannot…so they must escape to blow the two huge, holographic bubbles that are constantly being projected above and below the mobic scale.

This totals 11 dimensions in a layout of complementary space and time that is symmetrical across both bubbles. Having just 11 dimensions, not 12, is fortunate. It stops the propagation of more dimensions by jamming the master code's number lock. How? First, that uneven count of 3.5 dimensions per bubble flouts the Rule of 4. Next, as 3.5 dimensions per bubble, the 3 and the 5 are both prime numbers. Next, 3.5 dimensions per bubble × 2 bubbles = 7 dimensions, which is another prime number. Next, add in those 4 dimensions at the mobic scale to get a total of 11, which is yet another prime number!

The 3, 5, 7, and 11, plus that awkward parsing of half-dimensions…all of it departs so radically from the Rule of 4 that any higher order of ratio-building is halted. It specifically defeats analog number's built-in drive to proliferate more orders of dimensionality. That multi-phase number block ends it at last.

Why, you still insist? See, it's just too hard to establish more complex orders of analog, relational, polarized, dimensional space and time by incorporating ratios that use 3.5, 7, 11, or their multiples. In effect, the specific subsets that developed the 11 dimensions broke off the analog key in its analinear lock.

Smart move! If each ½D arrow really *had* counted as a full dimension, and thus if the universe actually *had* generated 12 dimensions…well, 12 is divisible by 2, 3, 4, and 6, so dimensionality might have burgeoned onward to become a cancerous growth in this universe. Proliferating dimensionality might have escalated like a greedy tumor before the universal mind ever matured enough to sculpt any suns or planets in this upper bubble, much less tinker up us. By establishing its layout of dimensionality on a foundation of so many prime numbers, the cosmegg rolled a winner at the craps table of creation. Hooray, for 3 verging into 5 and 7 come 11! What a big win!

6. Our universe is a Double Bubble attractor

All that is left now to do is adorn the gift package of dimensionality with that endless 8-bow of the tensor network. In each hourglass cell, the 8-loop extends across both bubbles to form a tensor network whose polarity continually oscillates between bubbles. Yet as it circulates, it never exactly retraces a previous line. It is always moving onward, always evolving the maturing universe itself.

This 8-looping dynamic recalls the iterating path that plays across both domains of a Lorenz attractor. The progression is 3D, nonlinear, and deterministic. Its looping path, which tilts much like a butterfly's wings, tracks a flow of chaos data across two domains. Its dynamical system uses three coupled, ordinary differential equations (ODEs) to emit the path of a continually emerging solution that evolves in a complex, non-repeating pattern. Edward Lorenz's 1963 equations for this dynamic describe a strange attractor.

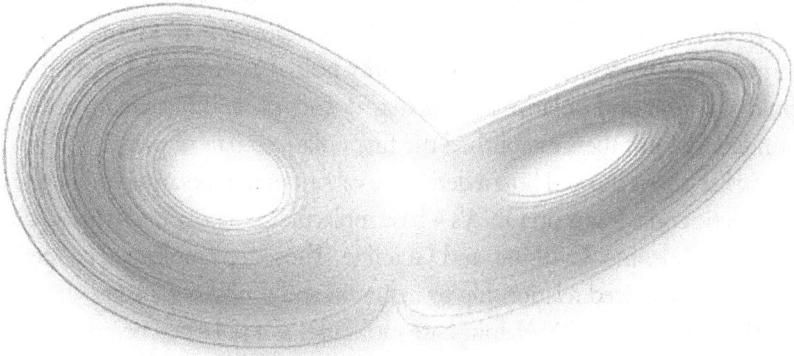

Lorenz attractor loop

This TOE declares our universe is a very special kind of strange attractor. Its two domains, basins, or 3D "butterfly wings" of oscillation are the two huge 3D bubbles with mirror-twin properties. One bubble holds 3D space and the arrow of time, while the other bubble holds 3D time and the arrow of space. Its dynamical system is the Double Bubble universe...or DB attractor.

The constant tensor network moving across both bubbles originates at the mobic scale. It derives from the mobic twist of a polarized 8-loop moving around both faces of each 2DD triangle in each pore, flashing to form a 3DD tetrahedron whose two volumes project an hourglass cell. And those 8-loops are projected, too. They ride along the corners of each hourglass's triangular funnel, appearing in either bubble as the arrow of *now* above or the arrow of *here* below.

Coupled together, the two 3D volumes act as its three ODEs (ordinary differential equations) that evolve the continually emergent solution of reality in each holographic bubble. This evolving solution of ongoing reality emerges everywhere in the upper bubble on time's omnipresent point of *now*. It also emerges everywhere in the lower bubble on space's omnipresent point of *here*.

A fractal has a basic, recognizable form, but its contents iterate in evolving, variable details. In each bubble, the iterating progression of events is 3D, nonlinear, and deterministic in its fractal form, yet also undetermined in its specific contents. In other words, each moment carries basic, recognizable fractal patterns that are deterministic, already set. We can, for instance, predict the general form of an established succession of basic patterns emerging in the constant *now* of our bubble...sunrise, moonset, food, government, work, leisure, sex, etc. But the pretty predictable form of events will hold variable specific contents. Often we can affect them by our conscious choices. We may even do it enough to alter basic fractal dynamics in future reality, for we live in an ever-emergent solution that constantly evolves. In fact, we *are* the ongoing solution.

7. The Rule of 4 jump-starts a co-chaos system

To think that it all started with "nothing" realizing itself as something: a number—0. Its *off*-again, *on*-again identity pulsed as binary 0-1. And that 0 could also split and polarize into analog -1 and +1, another polarity. Those foundational numbers established the first polarized pair of pairs.

Earlier books showed how a derivative variant of that polarized foursome is the genetic code. It sets up DNA's 4 base molecules as a polarized pair of pairs: **T**hymine, **A**denine, **C**ytosine, and **G**uanine. Those four molecules cooperate in a secure, polarized relationship to maintain and evolve organisms.

This chart shows DNA's four base molecules as a polarized pair of pairs:

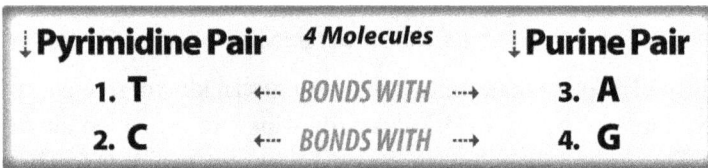

↓ **Pyrimidine Pair**	*4 Molecules*	↓ **Purine Pair**
1. **T**	←--- *BONDS WITH* ---→	3. **A**
2. **C**	←--- *BONDS WITH* ---→	4. **G**

Panel 1: the 4 DNA molecules cooperate vertically & horizontally

The genetic code's DNA and RNA can be shorthanded by I Ching math. This chart summarizes the math of the I Ching's polarized pair of pairs:

Yin-based ↓	*4 Bigrams*	**Yang**-based ↓
stable yin 1. ⚏	←--- *STABLE PAIR* ---→	stable yang 3. ⚍
changing yin 2. ⚎	←--- *CHANGING PAIR* ---→	changing yang 4. ⚌

Panel 2: the 4 bigrams cooperate vertically & horizontally

The genetic code and the I Ching are two latter-day variants templated from the master code. Its primal foursome are space, time, matter, and energy. They are a polarized pair of pairs: space-time and matter-energy. Like lesser variants, the four primals cooperate in a secure, polarized relationship to maintain and evolve the universe. Do you think that it wasn't intentional? Think again.

This chart summarizes the work order given to the four primals:

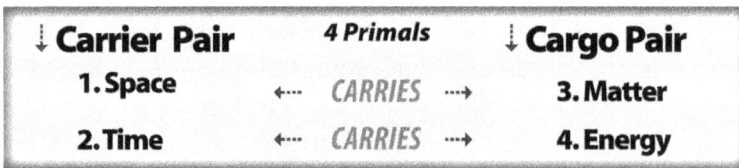

↓ **Carrier Pair**	*4 Primals*	↓ **Cargo Pair**
1. Space	←--- *CARRIES* ---→	3. Matter
2. Time	←--- *CARRIES* ---→	4. Energy

Panel 3: the 4 primals cooperate vertically & horizontally

Each primal pair is polarized to take on a fractal task in tandem with the other primal pair, and so all four primals have agreements that hold across both bubbles: These fractal agreements are simple but comprehensive:

- **1. Space & time are the carrier pair; matter & energy are the load pair.**
- **2. Space carries matter; time carries energy.**

Dimensionality used the stabilizing, polarizing Rule of 4 on a pulsing path to develop the 2DD triangle faced by 2D space and 2D time, and yes, dimensionality also rose to an even higher order when its 3DD tetrahedron projected D space and 3D time. But the attendant pair of half-dimensions started a prime number cascade that broke the analinear number lock and stopped dimensionality from turning into cancerous overgrowth. Instead, that tied a bow on dimensionality's dynamic stability. This fail-safe number security is at work in each master code variant. Result? We live, and our universe lives.

8. The Rule of 4 appears at every level

The Rule of 4 decreed by the polarized pair of pairs still holds sway for many objects, events, and conditions. We recognize its uncanny power at an instinctual level. We respond to it, all unaware that we are honoring the power of the Rule of 4 when we put four wheels on a car, set four stabilizing legs on a table, call a four-leaf clover lucky, or fear the four horsemen of the apocalypse riding toward us…from the four directions…on their four-legged horses.

The Rule of 4 is why a stubborn fourth point leaped out from the triangular plane at the mobic scale to generate the first tetrahedron. Like that influential fourth point, always the fourth leaf of a clover, fourth musician in a quartet, or fourth partner in a bridge game will stand out somehow in the foursome.

The 4's archetypal energy even turned a football team's quarterback into its star. In 1900, a new rule of the game set the Rule of 4 into the sport. Parke H. Davis explained it in his 1911 classic *Football–The American Intercollegiate Game*: "The man who first receives the ball from the snap-back shall be called the quarter-back."…"By creating the position of 'quarter-back,' football's founders created a man on the field who would stand out among equals…."

Before the T-formation came along in the 1940s, the whole offensive backfield was a legitimate threat. They all could run or pass the ball, and most teams used four offensive backs on every play: the fullback, two halfbacks, and a quarterback. In other words, they were a power-house of four offensive players. But the quarterback had a tweak of essential difference. He could also call the team's offensive plays. That made him stand out among the quartet.

That essential tweak of difference is why one musician in a rock band foursome will stand out to us. It is why the male Holy Trinity of Christianity needed that essential difference of female Mary. It is why the Hindu god Shiva, Roman god Neptune, and Babylonian demon Satan—take your pick—will hold a 3-pronged trident whose handle plunges in the opposite direction that is needed to probe a vital fourth condition of difference.

That essential difference of a fourth, eccentric, oddball, inventive factor is on display when Einstein draws the traditional signature of experiential spacetime in our upper bubble (courtesy of Hetemeel.com). In this trident's layout of the dimensional quartet, its essential difference is a handle of time thrusting downward in polarized opposition to 3D space.

Let Einstein show you the old spacetime signature

However, that traditional, four-pronged spacetime signature shows only one footprint of the universal Double Bubble bird. Its other footprint is a timespace signature that hides beyond the mobic scale in the mirror-twin bubble.

The dynamic of a trio calling up the existence of a different yet vital fourth factor or component occurs at all levels in this bubble, material or intangible. It exists even in your psyche, where it generates your shadow as a repressed fourth function, described in Jungian psychology, and mapped by the MBTI.

If one's shadowy, ignored, rebellious, even vile-seeming fourth function manages to become more integrated, civilized and refined, its different angle on life can even become worth its creative weight in gold. Always the essential difference of a fourth factor will rouse the quartet to do its best…or worst.

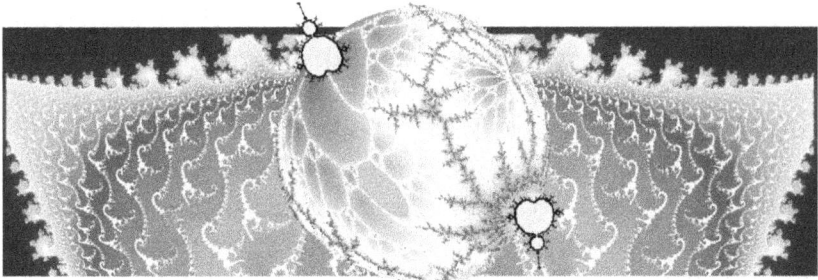

Chapter 16: The Stink Hole Prophecy

During deep-see diving in this universe, the more sentient beings I've met generally view our Earth as a stink hole of corruption and degradation. Why? Because humans not only treat each other badly so often, but they also so often project onto the larger universe the warlike, conflict-ridden traits of our own primate species with its posturing aggressions, fears, greed, and knee-jerk revulsions. Our current mindset and media usually portray any alien species as monstrous invaders ravenous to exploit this planet and destroy us.

Therefore, intelligent beings beyond this solar system consider it fortunate that gravitation holds us down to planet Earth physically.

We humans assume the only way to escape Earth's gravitational bonds is to fling our bodies upward in rocket ships. Some more intelligent beings in this universe know otherwise. Yes, gravitation imprisons our bodies, but not our minds. Until we recognize the other bubble and realize what it says about this universe's larger reality, gravity will lock our minds out of grand possibilities.

1. Visitors from beyond our planet

Here is what I see waiting for us as a possibility forming in the lower bubble's timescape. About 700 years from now, no more war on Earth at a governmental level. Given another 300 years or so, our new social maturity will eventually allow beings to start visiting many of us mentally from Proxima Centauri, the nearest star in a triple-star system called Alpha Centauri.

Consciousness is the only interstellar activity of Proxima Centaurians, but due to our own militant heritage, even their name miscues us to fear the invasion of centurians in our proximity. They have been patiently waiting for a time when they can appear to more of us—not physically but mentally—without destabilizing Earth's society. If they do come, they will visit us not in ships but in thought.

As we move beyond industrial blight to become simpler people, less complex in the sense of having fewer neurotic complexes, we will start to tap into their mental presence more readily and understand who they are as beings.

Many primitive humans have met them in the past, but it was generally experienced as the anonymous gift of a good idea or a calming sensation. They never tried to harm or scare humanity, but they found us too emotionally and culturally limited to understand who they are.

Elephants, dolphins, orangutans, and several other species already do grok them in a way that nearly all humans cannot nowadays—at least not without having an experience that sounds crazy to other humans and is perhaps diagnosed that way psychiatrically.

When I first met these beings, they conveyed to me that they are made of crystals. I thought it meant they were made of silica…that they were beings whose thought processing was due to some electrical action in a complex sand. Such mentation was hard for me to grasp, but I decided they meant they were basically hills of computer sand with giant brains. They seemed so wise and discussed things among themselves so thoroughly that I began to call them the Sandhedrin. It was a naming play on this Earth's Sanhedrin.

Later on, I discovered that my first encounter held a misunderstanding about their physical makeup that perhaps cushioned me to accept the even stranger form they actually do take. The Sandhedrin apparently thought it was a good joke that I at first assumed they were crystals made of sand, not ice…and hills, not vales…and one, not plural.

By the way, if all this is becoming too much for you to swallow, and you suspect I'm telling you what must be a hallucination, joke, or science-fiction gimmick, then call it whatever you choose, and let's get on with the account.

You see, the Sandhedrin live in the vales of a small ice-covered planet. They were created when mere hollows in the planet's surface abruptly filled with frozen atmosphere as the planet was kicked suddenly into an orbit quite far from its star. In that atmosphere abruptly chilled to near absolute zero, crystallizing events occurred that somehow allowed a thinking entity to form and evolve rather rapidly. It's a crystalline brain that covers the planet, and it quickly developed a facility for thought that exceeds our own. It is a pure mind-brain, calmer and more rational since it neither eats, drinks, kills, nor reproduces.

I did not suppose such an entity could actually exist, much less think. But I had at least a few ways of checking this out, so I can say that these details are as true as I can understand them…so take this chapter as you will, and deduce from it what you may.

Why do I sometimes refer to the Sandhedrin as "them" in the plural when they are actually a single, planetary brain? It's because each hollow is full of a thicker basin of mineral-laden ice in that ice-covered landscape, and each hollow holds its own subset of identity, so to speak. They are only five sentient vales, yet they are joined at the edges by an overcoat of frozen atmosphere that turns the small ice planet into a glittering sequin shining almost like a star even so far away from its sun.

These beings, the five-in-one, will not last as long as their planet. As their own sun, Proxima, drifts even farther from the other two suns (binary) that make up the Alpha Centauri star system, the resulting gravitational shift will paradoxically pull the Sandhedrin's ice planet into an orbit so close to their own sun that they will likely melt and die within 45,000 years.

Bound as the Sandhedrin are to their planet, their form, their constitution, the five-in-one do not propagate or reincarnate. Once melted, they will retreat into the memory of 3D time's great mind below the mobic scale, never again to emerge as material entities. They will finally live only in the lower bubble's huge unified mind, in the Grand Organizing Design that is more than all universes, and in the memory of the friends they have cultivated throughout the upper bubble, including me. I feel enriched to know them.

2. Head and heart go hand in hand

As I understand it, all life in this universe is here for a particular reason: to learn what God needs to know to become truly human, truly anteater, truly oak tree, truly Sandhedrin…whatever. Which sounds odd to me. You'd think it should be the other way around…that we're supposed to become more godlike.

But I'm told we are here so God can get in touch with itself at every level… and I hope "itself" doesn't sound clinically cold. All I intend to suggest is that the Grand Organizing Design is neither masculine nor feminine, but both and more—and we all act as the nerve-endings of God's own body, including the biggest nerve ending we have, the brain.

We are here to help the Grand Organizing Design (God for short) know, enjoy, and respond to each universe more completely. It created all of this to experience itself, develop itself—meaning all of us and everything else—in every way available, to find out what works best in each universe, and from that knowing, enable all to realize what is best for all to know: God.

In this way, God can grow, and so can we. We are God's being through doing, and this universe is about learning how and what and why to love. Many of us on Earth act like loveless wildings who scheme to get away with as much or more than goodwill allows. It may be the downfall of our human culture, and we will fall into God's arms…and find that we are God's arms.

But in discovering ourselves as an extension of God, as souls on a journey for the survival of the universe itself as a lover of God, we come to know why we are here, individually and collectively. In that way, God can know every possibility explored and rendered into a pathway to God. By that, we become what God wants for us in the divine plan that outlasts all of space and time and everything…everything except God. It renders everything finally one and as long-lasting as the Grand Organizing Design itself.

3. Life goes on

And what of us humans? Before too awfully long, we will find an energy source untapped at present, and its fuel will be water in minute quantities. No, it will not involve present-day nuclear power. It will be a form of energy that comes from manipulating water molecules in a certain way to gain maximum exposure of their inner workings at the quantum level. It will not afford any danger or risk to the environment or its users.

I cannot register or convey more than that because such an energy sounds so weirdly unlikely to me. Maybe that's why I cannot tune into how it works. Maybe that is for someone else to probe. So let us one day begin to consider a new way to tap into that quantum-level energy by first cultivating more consciousness. Sourcing the vast reservoir of mind in the huge bubble hidden below the mobic scale can offer us a new lease on life.

Eventually, we'll learn it is better to control the human population and become caretakers of this Earth, not its ransackers and destroyers. We'll decide to emphasize growing our souls, not our gross international product. It took over 2 million years for the world's human population to reach 1 billion. It took only 200 more years to reach 7.8 billion. After doubling the human population that currently exists globally, the vicissitudes of human population growing like a cancer unchecked on Earth will become so overwhelmingly obvious that we'll finally decide to, have to, must change our ways before it's too late for Gaia to recover and grow healthy again.

Eventually, our population will likely stabilize again at about half of the 7.8 billion that now exists. There will be great factory farms of a very different kind from what we now mean by that term. They will not be full of penned animals being bred for slaughter. These farms will instead be full of actual factories, mostly mechanized and tended by just a few humans in rotation.

As for animal farms, large ruminants will someday be allowed to roam in great herds such as the buffalo once did in North America, and as we presently see with sheep in Australia. For a time, from them will come the meat eaten by humans, dogs, cats, and other care-dependant animals, as well as the hides, wool, tallow, lanolin, and such that cows, sheep, and goats provide.

We'll still eat fish and chicken for a long time, but we will mostly cease to eat pigs eventually because they are so smart, the bright but handless victims of our obliviousness. We'll also keep and eat sheep and cows much longer, partly because they would rather be many than few, and also because it heartens us to know they are no longer warehoused and abused in the manner of livestock today, who often never even get to enjoy their own lives before dying.

Of course, for a time, we'll want some animal products that can only be obtained by maintaining herds. But we will also keep animals to remind ourselves that we are animals in human form. We are omnivores by heritage, with a bloody past that is still evident in the very shape of our canine teeth. Ignoring our genetic past denies the bloody shadow hiding in the human mind. It is subdued by centuries of social decorum, yet it still often displays how very temptable, vulnerable, and "primate-ive" our relic instincts are, despite the increasing age and wisdom of our species.

We will raise goats for a long time. We will keep them as our main dairy animal, mainly because due to their small bodies, kid-nursing rates, and contrary natures, we cannot exploit them in the way that cows permit. We will pride ourselves on our friendships with goats. Their sturdy survival values will teach us much about resilience.

Eventually, we'll develop artificial meats that are not only delicious but also provide all of meat's benefits, including the saturated animal fat and vitamins A, B, D, iron, zinc, selenium, etc. We'll decide that animals are not to eat, but to enjoy. Most of the animals that we now call domestic—horses, camels, cows, chickens, pigs—will live in small zoos in the towns, where people will pride themselves on how well they care for their animals kept not for food but for companionship, so that slaughter for food becomes only a dim racial memory for most.

Many in our current urbanized populations have almost lost touch with unsensationalized sensation, but in time, many people will begin to opt to learn from animals when they realize what animals have to teach us about the zesty common sense that resides in the natural body. As we manage to become kinder to animals, we also become kinder to ourselves, to each other.

We'll have more time to explore the habitats of animals in the wild, and we will create more wild for them to live in. Vast herds of wild and nearly wild herbivores will e eventually roam much of Africa again, and carnivores will be allowed to stalk them in the ancient way. Ditto for each continent, except Eurasia. It will slowly become home base for humanity, the place where we populate in sustainable cities.

But we'll also travel often and widely on every continent to enjoy nature and explore ways of communicating with animals, discovering they are more intelligent than we presently suppose. They will help us in ways that we've not yet considered, nor allowed ourselves to imagine in that time when we merely ate them or used them or hunted them for sport, rather than getting to know them by their gestures and in the silence of the mind. As the 17th-century poet Richard Crashaw said:

> Eyes are vocal, tears have tongues,
> And there are words not made with lungs.

We will find that animals converse with each other in physical nonverbals, and even mentally in ways that our ego development among humans has lost touch with, but which we'll need to regain if we are going to join in the universe-wide communication of all life at that deep level. We will find, for instance, that elephants and dolphins already honor the ice-minds of the Proxima Centaurians when we humans do not, for we expect math equations from visiting aliens, whereas the Sandhedrin actually offer soul wisdom, finding it more valuable than formulas for a technology they do not have and cannot use.

But they do know a kind of math quite well, and philosophy and histories of all sorts. They can tell stories of many worlds and ways, and do it well, with great humor and pithy detail, for I have heard some of them. They are more impassive in their approach to life than most humans or other beings I have met, so their humor is more deadpan in the telling.

By traveling the universe mentally, someday we will discover what the Sandhedrin already know—that most beings prefer to sustain and enrich the world they are native to rather than maim or destroy their own world while seeking to explore, exploit, or escape to another. As we primates come to cherish what we already have, we'll find that other animals here become our best allies in learning how to live well in this possible Eden glimpsed over the horizon.

More than that about the animals, I cannot say, except to add that in the possible Eden peeking over the far horizon ahead, we will come to understand that although we once distorted our animal natures into urban nightmares, we can redeem ourselves someday to join the larger community beyond this Earth.

When we realize that our thoughts can transmit just about instantly to a point anywhere in the upper bubble merely by bouncing it there via the tachyon cloudbank in the lower bubble, we will start to access foreign friendships for advice, consolation, and understanding of all kinds. Our psychological skills here will vastly increase as we gain more understanding of our own psyches and how the personal unconscious flows into the vast sentient pool where universal mind holds the collective experience of all souls on its many worlds.

We have limited ourselves to a rather narrow human view of what creativity means thus far, but the deep, endless wellhead of universal mind taps into an ocean of inspiration. From it, we can draw up visions for productivity of novel kinds that we have never before imagined.

As we communicate with other races beyond our star, our galaxy, our sector, the histories of many defunct war-like planets will become common knowledge to us,. There we will discover that war itself is an archaic social form cultivated by those species who have generally managed to kill themselves off, and perhaps even their planet, before ever colonizing beyond it.

4. Optimum size in people

Some souls are small in stature. Some are large. On Earth, we do not often think of people as big or little in this particular way. We mostly consider their physical or social or ego or financial size. But what of spiritual size? If that were measured, it would indeed be remarkable to see. Some human giants of ego, wealth, or social clout would be tiny in stature on the spiritual scale. Others tiny in worldly influence would be huge in worth before God's eyes.

God, I am told, is not an equal-opportunity creator. He gives some every advantage and some nearly none. The result is what God is interested in. How much can you do with what you've got? Can you parlay it into more spiritual stature…and more? How evolved is your soul going to be when it leaves your current body, and what are you setting it up for when you come back in again—as a hero of the soul, a slacker adrift, a villain jacked on perversity?

That's what God is interested in, I have found in the great mind below the mobic scale. It holds the spiritual accounts and lets souls accrue wealth as they will. When a human physical container wears out, its soul enters a temporary waystation between lives (Catholics call it *purgatory* before heaven; Tibetans call it the *bardo* state before reincarnation), and it usually emerges again in a new material body to pass yet again through this human hotbed of soul growth.

As we humans individually grow in size as souls, so does our collective human soul identity. Likewise, as the universe itself moves from incarnation to incarnation, it grows in self-realization. This means that as we become more than we were, both individually and collectively, we enhance the universe as a whole.

If our universe lives in God's mind and heart (and it does, for it knew how to access God even while transmuting a 0 of nothing into the wonder of all that is our universe), then so will the Sandhedrin, and so will we, and so will all the other species. We're carried forward in the consciousness of the whole.

This universe tries to make its life a joyful process of becoming more and more fully realized. How is that possible? I do not know exactly, but I perceive that it is possible to improve upon what is. When I look around at the suffering, the privation, the injustice in this world, and from what I've seen playing out in the species that interweave the story of this universe's ongoing reality, I realize we are all ready and waiting for a cue to act better.

History shows that we've already evolved our ethics a lot. Cannibalism, for instance, was widely accepted over many millennia. Even in the 21st century, that ritual is still practiced furtively in a few sub-cultures, in shock exhibitionism, or in an infrequent starvation case...but cannibalism is not the social norm.

Slavery was once an accepted economic "necessity" in most cultures around the world, and for a few, even in the 21st century. But now, chattel slavery is outlawed in nations worldwide, the last being Mauritania in 2007. Regrettably, vulnerable people still get bought and sold in human trafficking, but such behavior is no longer condoned as acceptable in any recognized nation.

War is still an accepted norm for most nations, but I look toward the day when we realize that we are warring against ourselves, just as we've already learned that we were eating ourselves, enslaving ourselves, buying and selling ourselves. I look toward the day when we can treat each other as friends of all stripes, visible and invisible, who know that God created us to know each other and this universe, and by loving it and each other, to know the divine.

In time we will finally have large amounts of nearly wild space, and within that natural landscape, yes, many small towns that are sprinkled like active basophils to guard against inflammation in the body of Gaia.

Traveling companies will circulate among the small towns and slightly larger cities like wandering musicians and players of yore, but they will be well-known from a sort of television where every community will have its own broadcaster to offer the best of its community to the world at large.

Due to so much techno-automation, we will treasure specialty stores, small manufacturers, cooks, craftsmen, and artists of all kinds in a truly niche market on a global scale. People will still work to create and even do it gladly, occupying perhaps 20 or so hours a month on average, in a satisfaction born of social commitment, not money-making, since power is almost free and time becomes more flexible as people gain in equitable wealth. In this way, the many gain from the relinquishment of great wealth hoarded by a few.

A few custom cities will specialize in impressive buildings, entertainment centers, and stadiums. They will sit spotted here and there in the wild terrain like islands of the urbanized past, memorial amusement parks reminding us of some joyful extremes of human extravagance. Perhaps one will look a bit like this...

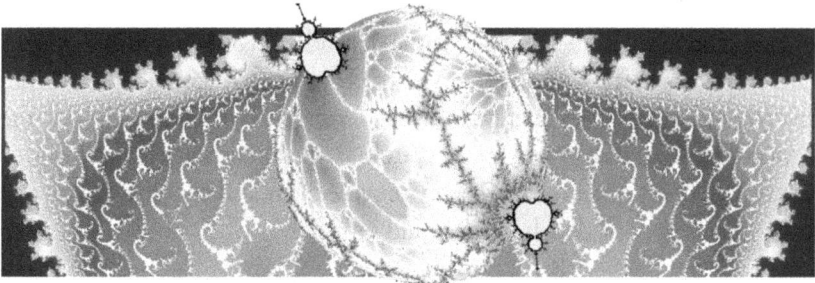

Chapter 17: The Pretty PPCCL

1. Mobic mirroring for the two bubbles

This chapter summarizes some key scientific aspects of the Double Bubble TOE. It describes a fundamental symmetry for our universe that provides mobic reverse-mirroring across both bubbles for the four primals of space, time, matter, and energy. The two primal carriers, space and time dimensionality, manifest in opposite, complementary ways across bubbles, and their two primal loads, matter and energy, must fit their carrier's constraints. The resulting general invariance in physical laws produces the PPCCL covariance between bubbles that represents a fundamental conservation law in physics.

To visualize this reversing-mirror symmetry in familiar terms, imagine with a child's mind that you stand before a big wall mirror. You see what looks like your twin there in the mirror. What you see is a flip-flopped image that reverses your own reality in a left/right way. The hands of your twin's wall clock sit at 10:00, but on the wall behind you, the hands of your own clock sit at 2:00.

You know that physics declares your body is made of matter, but as you look into the mirror, your twin insists that she's not a material girl...she is made of antimatter. According to science fiction plots, you two mirror-reversed twins must never meet and shake hands because as physical doppelgangers, the pair of you are made of oppositely polarized substances. Upon contact, you would both go KABLOOIE! and cancel each other out.

The Double Bubble twins also exhibit a mirror reversal. To see it, we must move the mirror from the wall to the floor of the world, so to speak...below the scale where matter and energy emerge at the quantum scale's known physical limit. We must go on down to the ultra-tiny mobic scale at the membrane interface between both bubbles. This is where space and time emerge. Each pore of this interface holds a mobic, polarizing warp that rotates between both bubbles. Its dynamic provides some aspects of a Mobius band and some aspects of a Lorenz attractor, so we elide it all into the *mactor* dynamic.

We stand in the interface and observe that in each bubble, the four primals of space, time, matter, and energy operate as a polarized pair of pairs. They are polarized such space and time act as the dependable container forms. Matter and energy act as the evolving cargo loads. Thus in each bubble, space carries matter, and time carries energy. In fractal terms, space and time provide the dependable form that gives a fractal pattern its distinguishing characteristic or shape. By contrast, matter and energy provide the fractal contents that iterate with varying specific details in space and time.

2. Space & time—the two universal carriers

First, let's consider the two universal carriers: *space and time.* How do they originate? The cosmegg's nothing-identity becomes aware of itself as something, the number 0. It alternates pulses of nonbeing and being (0 and 1) to make a 0DD point. Another beat in a slightly different locale turns that point into a 1DD line of tension polarized by space and time (+1 and -1). Then a third beat in a new location evolves it into a plane, a 2DD triangle faced by 2D space and 2D time. The cosmegg now has 4 dimensions operating at the mobic scale in a shared "equi-dimensionality" that is possible only at this scale.

A fourth beat in a new location turns that 2DD triangle into *four* 2DD triangles that shape a 3DD tetrahedron. The mobic scale can contain the quick-shifting, planar surfaces of all four 2DD triangles sketching the sides of a 3DD tetrahedron…but not its two volumes of 3D space and 3D time.

The outer and inner volumes of the 3DD tetrahedron project above and below the mobic scale as an hourglass cell stretching to the universal limits. This is a single hourglass cell of 3D space and 3D time with a wasp-waist interface at the middle. With that success, the cosmegg instantly replicates the single cell into many hourglass cells. They merge their volumes holographically into two huge bubbles—whoosh!—here is what science calls inflation.

3. Mobic mirroring for space & time

Physicist Eugene Wigner warned that the standard model's view of dimensional reversal is shortsighted and simplistic. Its reversal only postulates a time arrow that would go backward instead of forward, and only theoretically, since that never happens in real life. As for space, it provides no reversal at all.

However, in this TOE, a true symmetry of space and time exists across both bubbles. In the Double Bubble, time does not just go backward and forward, leaving space with no reversal. Instead, the upper bubble's 3D space and ½D time arrow flip-flop into their true complement: the lower bubble's 3D time and ½D space arrow. Thus in each bubble, one carrier can expand its load in three contiguous dimensions while its companion carrier must restrict its load to a mere ½D arrow of dimensionality.

As a result, the Double Bubble universe exhibits a mirror-reversed layout for space and time across bubbles. The upper bubble's spacetime trident is a 3D space fork with the ½D time arrow as its handle. The lower bubble's timespace trident is a 3D time fork with the ½D space arrow as its handle. This reverse-mirroring provides a more viable dimensional format than what's found in current standard physics, which does not recognize the lower bubble.

Both bubbles are joined by the membrane interface of ultra-tiny mobic pores. Those pores project all the hourglass cells that merge into two holographic bubbles. The mobic pores provide a dynamic that automatically mirror-reverses the upper bubble's spacetime layout into the lower bubble's timespace layout. It provides a balanced distribution of dimensionality across both bubbles—3.5 dimensions each—with total parity between bubbles in their number of dimensions. In effect, this layout steadies the Double Bubble bird by giving it two feet to stand on. Standard physics cannot do that.

The two huge bubbles of dimensionality in this holographic universe have reciprocal properties that manifest in opposite, complementary ways. The upper bubble has contiguous 3D space and the ubiquitous arrow of ½D time. Meanwhile, the lower bubble has contiguous 3D time and the ubiquitous arrow of ½D space. Thus, each bubble has two carriers: a 3D volume and a ½D arrow. Tally it up: the two bubbles hold 7 dimensions, plus the mobic scale interface holds 4 more dimensions. Result: 7 + 4 = 11 dimensions in all.

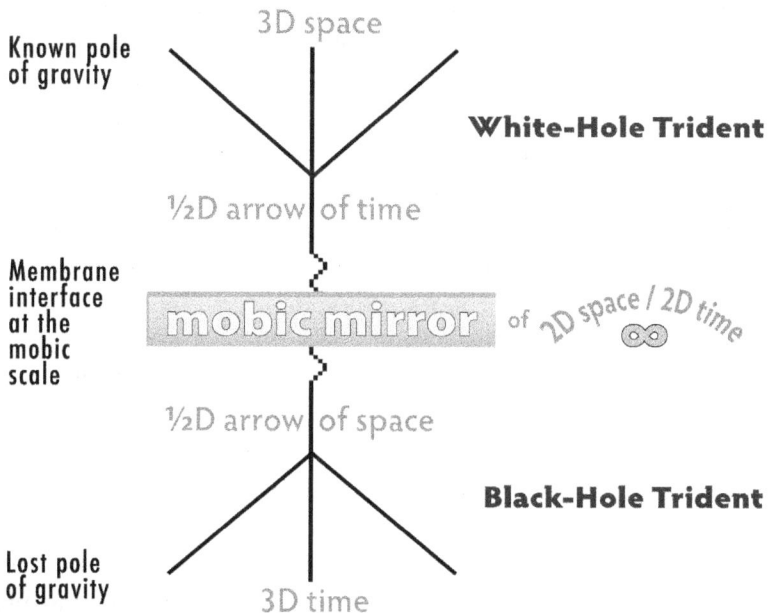

The Double Bubble's mobic mirror of its space-time footprints

4. Matter & energy—the two universal cargoes

We have considered space and time, the universe's two dependable carrier forms emerging at the mobic scale. Now let's consider matter and energy, the two fluxing cargoes emerging at the quantum scale. Matter and energy contents must fit their carrier constraints and iterate in changing specific details over space and time. Thus, space and time provide the recognizable, dependable fractal forms. Matter and energy provide the iteration of varying fractal contents.

What generated the two cargoes? Any pulsing left over from making the space and time lattice was polarized residue that got shunted into the two holographic bubbles and repurposed there. Half of the debris was polarized to go into our upper bubble's layout and become mattergy; the other half was polarized to enter the other bubble's dimensionality and become antimattergy.

Above the quantum scale, the 3D spatial carrier manifests its matter load as particles that can bond into pop-out, 3D objects held together by energy carried in the ½D time arrows. However, those arrows restrict the energy load so much that it can only manifest as waves coiled in the arrows, traveling only ahead and only up to the speed of light. This arrangement lets the particle-wave contents group into pop-out atoms and molecules in 3D space, iterating their changes over ½D time. Macro-molecules organize in supportive environments throughout the upper bubble's spatial volume, accommodating many diverse little organisms with portable mini-minds, like us here on planet Earth.

Meanwhile, half of the construction debris was shunted by polarity down below the mobic scale into the lower bubble. Down there, its ubiquitous ½D arrow of space squeezed the would-be particles of antimatter with such intense spatial pressure that they all converted into energy, for $E = mc^2$ holds true universally. The antimatter was converted into zippy tachyonic energy with imaginary mass (imaginary in the number sense, that is).

With a happy result! Converting antimatter into energy lets it escape the confines of ½D space to travel in 3D time as elaborate patterns of tachyon particle-waves going only above the speed of light[2]. This tachyonic energy is organized into a single, huge mind in the 3D time bubble that acts as its flex-form skull. That great mind spread throughout the lower bubble exists in the continuous, omnipresent *here* of ½D space across all contiguous 3D time. The uniform locality of constant "here" in that bubble allows that great mind to transmit data throughout its tachyonic cloud via literal "instant messaging."

And it can also communicate to the upper bubble's 3D space via the mobic membrane, transmitting data instantly between locales. Thus the lower bubble's constellations of energy patterns spread throughout 3D time in constant *here* are far more sophisticated than the meager displays of energy that exist in our upper bubble's 3D space, confined as our bubble's energy is to perpetual *now* moving on the thin arrow of ½D time.

This topic also references ⊞ **Question 16: How can two particles communicate faster than the speed of light?** Instantaneous data transmission at the interface between bubbles causes the phenomenon that current physics calls quantum entanglement, nonlocality, or action at a distance. Current science cannot yet explain how two objects separated in 3D space can communicate instantly to update their relationship without a known intermediary, agency, or mechanism. This instant communication happens because universal data is coordinated everywhere at the mobic scale.

Since our bubble and the other bubble constantly communicate via polarized pulsing at the mobic-scale membrane, we get the instant benefit of that huge unified mind. Some minimize it as Mother Nature. Some maximize it as God. But our universe's mind is not God. God is greater still.

Above the quantum scale, along with emergent gravitation, this bubble also has a trio of lesser energy forces. Each force has two poles, and the energy only moves ahead on the arrow of time, supporting the upper bubble's diverse little organisms with relatively slow mini-minds clustered here and there in 3D space. The organisms try to surmount the ½D time stricture by developing instincts that can track patterns in time's larger scope. They sense mating season, migration cues, birth pangs requiring a safe haven for delivery.

More consciousness has let some organisms use time well enough to enhance life beyond mere survival. Some species in this upper bubble (like us, for example) even track on calendars the eclipses and high tides, the sowing and harvesting, even birthdays and holidays. By now, our species can measure time with atomic clocks and light-years. Nevertheless, despite our increasing advances in technology, time still frog-marches us all ahead, never backward.

The universal mind in the lower bubble continually tracks at the membrane interface all events in the upper bubble. It makes many tweaks to support this bubble's life, yet it also constantly incorporates whatever occurs due to those little organisms' measure of free will. To keep the universe going, it must make both accounts jibe in a double-entry bookkeeping at the mobic scale.

Unfortunately, we humans, with our limited perspective and our measure of free will, tend to falsify reality. Mainly we do it unconsciously because of our compartmentalized and biased perceptions. But we also do it with conscious effort, in lies that feed intentional distortions into the mix. Such distortions only make it harder for our species to understand what is really going on.

5. Mobic mirroring for matter & energy

Current science speculates that theoretically speaking, matter and antimatter versions of you could exist as doppelgangers with opposite charges. Upon meeting and shaking hands, you would both automatically cancel each other out. You'd both explode and disappear, releasing lots of energy.

This TOE disagrees. It declares that the reversed charges of matter and antimatter indicate something radically larger in scope. In the Double Bubble universe, a meeting between two large-scale entities of matter and antimatter such as you and your twin cannot possibly occur. The universal parameters do not allow it. Complex systems such as yourself and that twin looking at you from the mirror could never meet as physical doppelgangers in a head-on confrontation and go BOOM!...

...because you two are reciprocal kinds of matter and energy carried in reciprocal containers of space and time. You are molecular matter-based, but she is tachyonic energy-based. This means you two function in profoundly opposite ways in the reciprocal dimensionality on either side of the mobic scale.

To be specific, your body's 3D bulk is made of tardyon matter particles located in 3D space, glued together by the slow energy waves coiled in ½D time. However, your complementary twin's body in that other bubble is not so neatly limned into a material shape. Her antimatter has been compress-converted by that bubble's narrow gauge of ½D space into super-fast tachyon energy with particles of imaginary number mass.

Your reciprocal twin is a 3D constellation of tachyonic energy, a cobweb of resonant, buzzing connectivity vibrating among all the many other patterns vibrating throughout the lower bubble's contiguous 3D time. Her shape exists as a cloud of swift energy patterned with your identity data, located somewhere in the omnipresent *here* of that vast unified mind.

You two mirror twins cannot possibly inhabit the same space and time, yet you are complementary aspects of a shared identity. In a sense, you often do meet. You do it nightly when you sleep. In dreamtime, your ego identity encounters hidden patterns that are evident to your larger, unconscious self. Its symbolic scenarios embed the dream "I" of your ego into teaching scenarios.

In a dream, your ego "I" walks around experiencing things that symbolize issues in your waking life. Often your conscious ego has contrived to botch or ignore those issues, leaving them unresolved in your life. Your dream will depict in symbolic pantomime (usually exaggerated, often nonverbal) how various choices can nourish or stunt the growth of your ongoing soul, whose lifetimes are stored in a constellation of "you essence" in that other bubble's tachyonic cloud.

Your dreams are a gift from universal mind, dream mail sent to you because it cares about you. The dream's larger perspective tries to inform and adjust the ego "I" walking around in the dream—your sense of identity that experiences things in the dream events. Its plotline sets your ego "I" into little parables that seek to cue your waking identity into a new realization, attitude, or action.

After a dream, your ego may become aware of what your soul pattern nestled in universal mind wants your ego to wake up to. That greater perspective in 3D time sees and tries to improve upon the dynamic of what's happening in your present life's situation here in the upper bubble, a dynamic that stretches well beyond your own ego's recognition.

If you have recurring nightmares, they are trying to get your attention by monotonously hitting you upside the head with a metaphorical 2 by 4. The more extreme and frequent the nightmare, the harder it is trying to get you to alter your attitude, stance, behavior, or approach to life somehow. It hopes to warn you about a physical or mental harm, or it tries to show you how to deal with a harm already done. But it usually speaks in symbols, not data points.

Dreams offer you holistic connection with transcendent meaning. If you are attuned to dreams, either by natural gift or careful cultivation, you can pick up on the subliminal cues they contain. Dreams can heal old wounds by reframing a trauma in your past. They can help you deal with a present-day quandary. They can coax you to alter a stance well enough to find a better route into a custom-fit future. But if you are not attuned, your waking mind will probably just trash each dream along with its undeciphered contents.

Please realize that your lower-bubble twin facing you in the mirror knows things beyond your own ego's values and perceptions, for she exists in 3D time. She offers you a corroboration when appropriate, a corrective where necessary. She is both luminous and numinous, full of memory and brimming with future potential that is not yet actualized. From a physics point of view, she is just a tachyonic energy field in the databank of universal mind. From a philosophical viewpoint, she is your soul nestled in the universal mind.

Both she, made of numinous 3D energy dwelling in expansive time, and you, her matter twin perched up here in the latitude of 3D space—you both exist in the Double Bubble. You both are entwined inside this superb universal body that is polished at the mobic scale by tensile beats of gravitation striking constant sparks of adamantine illumination in the universal mind.

6. Mobic mirroring for gravity & antigravity

Yes, I know this TOE goes counter to current physics, which claims that gravitation is a monopolar force. The universe's only monopolar force. Really? As if a force could have just one pole without its polarized opposite?

Come now! Polarity insists on 2-ness!

But not only does current physics view gravitation as a monopolar force in a 3D space bubble that it calls the whole universe. It also claims the "sole" gravitational pole is positive. However, this TOE declares that our universe has two bubbles with two poles of gravitation, one in each bubble. It also insists that the gravity pole here in this upper bubble is the receptive, negative yin pole.

Why do I say that? At inflation, the first force of gravitation shunted one gravity pole into each conjoined mirror-twin bubble. Our upper bubble got the pole whose dynamic is the receiver of action. Its receptive yin pole of negative energy attracts mattergy. That's why apples fall to the ground, not upward. Why mass clumps together with other mass. Why light waves bend around a planet. All those events demonstrate that our gravitational pole is the receptive, negative, yin pole. Our pole attracts matter, bends light, clumps things together.

That other bubble got the positive, assertive yang pole of gravitation. Its yang pole of positive energy assertively repels mattergy. It forces the tachyon particle-waves with imaginary-number mass to separate, stretching and expanding the 3D time flexnet. This action enlarges the lower bubble's huge brain capacity existing in omnipresent *right here* without any *over there*.

Thus gravitational force presents in complementary ways within the two bubbles. If you consider how both gravitational poles act across the Double Bubble universe, you realize the upper bubble has the *negative* pole of gravitation, attracting objects to it, while the lower bubble has the *positive* pole of gravitation, pushing objects away. It's the only way both poles can function.

And here's another downer for sci-fi plots: the Double Bubble's gravitational setup means our science cannot make true antigravity belts. Don't confuse true antigravity with various devices that use voltage, leverage, or ionic wind to mimic the effect of levitation that is "freed of gravity." They do not really free mattergy from gravitation; they merely use some means to counter the pull of its attraction temporarily, and hey, you can do that whenever you jump.

Antigravity is not going to waste, however. The lower bubble uses antigravity to push the tachyon energy patterns outward, expanding the balloon of 3D time for the enlarging universal mind. Antigravity's "push" dynamic gives those brain waves plenty of room to elaborate on the omnipresent arrow of ½D *here*.

Now I ask you: how did physics get so provincial about gravitation that it hasn't noticed the bigger picture? To me, the definition of gravity is now tied so strongly to velocity and acceleration that science and technology often act like the only important issue is, "Do gravity, velocity, and acceleration all have the same or opposite signs in a problem?"

And to me, current physics seems muddled about polarity in general... on the topics of dark matter, dark energy, and neutrinos, for instance. In the

Provenance episode of the *Numb3rs* TV series, (actor) physicist Larry remarked, "Some of the greatest errors in cosmology come not from poor math, but from poor assumptions." This TOE suggests that cosmology's main poor assumption is simply not recognizing that our bubble has a symbiotic mirror-twin, a conjoined bubble of complementary properties. Not noticing that both twins exist has caused physics to promulgate some errors of misconception regarding time, gravitation, antimatter, dark matter, and dark energy, for instance.

7. CPT symmetry vs. STEM symmetry

The Double Bubble TOE proposes that a foundational STEM symmetry exists in our universe. Its symmetry is far more comprehensive than standard CPT symmetry. STEM symmetry reverses the properties of space and time into the true complement of each other. It does likewise for the properties of matter and energy. Its two bubbles exhibit mirror-reversing symmetry for the four primals of **S**pace, **T**ime, **E**nergy, and **M**atter—or **STEM** symmetry.

❖

COMPARE CPT SYMMETRY WITH STEM SYMMETRY...

1. In CPT symmetry, C stands for Charge reversal of particle-waves. Charge reversal in physics says reversing the electrical charge on a matter particle signifies that it is now antimatter. The Double Bubble TOE agrees with this for any antimatter created here in the upper bubble.

But it identifies charge more profoundly, saying that our universe's first great symmetry-breaking occurred when inflation split the evolving cosmegg into two mirror-twin, conjoined 3D bubbles, automatically shunting an emergent pole of gravitation into each bubble, and thus also automatically sorting and shunting the components for original matter and antimatter into their two polarized bubbles.

In the upper bubble, our 3D space and ½D time hold tardyon particle-waves of original matter and slow energy traveling only up to the speed of light. Matter up here can construct large structures that are held together by energy waves coiled in ½D time arrows moving in the upper half of a tensor network that causes our constant arrow of ½D time to shoot ahead in omnipresent *now*.

In the lower bubble, its omnipresent arrow of ½D space long ago converted the original antimatter into tachyon particle-waves with imaginary-number mass and swift energy traveling only above the speed of light[2] in 3D time. Its quick energy powered up and evolved the large, complex patterns of a unified, universal mind spread throughout its 3D time, continually updating via the lower half of a tensor network that causes its arrow of ½D space to shoot ahead in omnipresent *here*.

2. In CPT symmetry, P stands for Parity of particle-waves. Parity reversal in physics says that reversing the spin of a matter particle so that it's both upside down and backwards in space signifies that it is now antimatter. This TOE agrees regarding any antimatter that is created in this upper bubble.

However, the Double Bubble STEM offers a more profound symmetry. A matter particle is polarized to occupy 3D space, and its energy waves are polarized to ride coiled in arrows of ½D time. Conversely, an antimatter particle of imaginary-number mass is polarized to occupy ½D space; and its energy waves are polarized to organize in constellating patterns throughout 3D time. That's why antimatter cannot last long in the wrong polarity of this upper bubble. If antimatter is created in the wrong specs of this upper bubble, it seeks annihilation by finding a matter twin to cancel each other out, disappearing in a release of energy.

3. In CPT symmetry, T stands for Time reversal of particle-waves. Physics claims that time reversal would turn particles into antiparticles ... and vice versa. But this TOE suggests a more profound symmetry. Time's omnipresent arrow in the upper bubble is the upper half of a huge tensor network that is 8-looping across both bubbles. Likewise, space's omnipresent arrow in the lower bubble is the other half of that 8-looping tensor network.

The tensor network's ubiquitous 8-path circulating across both bubbles creates a *double-torus* effect, and reversing its toroidal motion simply cannot occur. Motion only goes forward on the 8-loop of its double orbit across both bubbles. This reads as the constant *now* of ½D time moving forward in this bubble. Likewise, it reads as the constant *here* of ½D space also moving forward in that other bubble.

Reversing the tensor network's toroidal motion cannot occur, so in our upper bubble, a particle of antimatter cannot go backward in time. However, it can *appear* to. What? How? In its push against this bubble's time specs, it in effect pauses in time until a matter particle catches up with it, and they mutually annihilate.

THE SUMMARY FOR STEM SYMMETRY...

The Double Bubble's dimensionality has a two-trident layout that points above and below the mobic scale. The space and time tridents are counterbalanced carriers of their matter and energy cargoes when noted across both bubbles.

In each bubble, its space and time act as the two carriers. They impose **P**arity and **C**harge requirements upon their two cargos, matter and energy. These counterbalancing specifications across bubbles enforce a profound mirror-symmetry for all four primals in the Double Bubble universe.

A1. Properties of the two carriers—3D space volume expands in the upper bubble, but 3D time volume expands in the lower bubble. Stretching across both bubbles from the mobic-scale interface is the tensor network, a double torus of 8-loops. Its upper half reads as the ubiquitous arrow of ½D time in the upper bubble; its lower half reads as the ubiquitous arrow of ½D space in the lower bubble.

A2. Space & Time have P & C balance—across both bubbles, parity in the dimensional layout gives each bubble a trident of 3.5 dimensions, with polarized charge properties that are mirror-reversed across bubbles. This flip-flops the layout of the spacetime trident above into that of the timespace trident below. Thus the parity and charge for space and time balance across bubbles.

B1. Properties of the two cargoes—the upper bubble's 3D space carries its mass as tardyon particles bonded into complex 3D structures, but the lower bubble's ½D space carries its anti-mass as tachyon particles of imaginary number mass.

The upper bubble carries its energy as tardyon waves coiled in its ½D time arrows, but the lower bubble carries its energy as tachyon waves constellating into complex patterns in 3D time. Thus the upper bubble's lattice of 3D space supports complex structures of 3D matter, while the lower bubble's flexnet of 3D time supports complex patterns of 3D energy.

B2. *Matter & Energy have P & C balance*—due to the mirror-reversed properties of the space-time carriers across bubbles, their iterating cargoes of matter-energy also exhibit mirror-reversed traits.

In this way, the Double Bubble's STEM symmetry indicates a covariance for space, time, energy, and matter across both bubbles. Current physics cannot offer so much symmetry.

❖

Double Bubble TOE symmetry is far more comprehensive than standard CPT symmetry. STEM symmetry reverses the properties of space and time into the true complement of each other. It does likewise for the properties of matter and energy. Thus it provides mobic-mirror symmetry by reversing the properties and charge for all four primals across both bubbles.

In this way, the Double Bubble TOE not only accommodates the CPT particle-wave symmetry of current physics, but also absorbs it into the more comprehensive order of STEM symmetry across bubbles.

So much correlation for the universe's primal polarized pair of pairs reinforces the myriad relationships that make the Double Bubble hologram strong, dependable, and failsafe. It securely maintains the universal body and mind, while also permitting an ongoing evolution of the contents within its two bubbles, and it even allows some leeway for free will and error slippage.

8. The STEM symmetry provides PPCCL covariance

PPCCL

P...**P**arity between space and time
P...**P**arity between matter and energy
C...**C**harge reversal for space and time
C...**C**harge reversal for matter and energy
L...**L**ight speed conjugation of matter and energy

The PPCCL of our TOE interprets certain mathematical symmetries in physical properties to reveal an important new symmetry in physics. It follows the example of Einstein, who developed his mathematical concept of spacetime symmetries into a physical principle. He said the main idea behind his theory of general relativity was *general covariance*. It dealt with an invariance of physical laws under arbitrary differentiable coordinate transformations.

How important are invariance and covariance in physics? In 1918, mathematician Emmy Noether showed that for every symmetry in physics, there is a corresponding conservation law. Transformations in spacetime represent shifts between the reference frames of different observers, and invariance under a transformation represents a fundamental conservation law.

Identifying invariance in special relativity was a great step forward, yet according to Eugene Wigner and others, the standard view of space and time dimensionality remains flawed, for it is not truly balanced or comprehensive. Wigner warned repeatedly of its incompleteness in *Invariance and Physical Theory*: "Again it may be well to remember that this invariance may have limitations."

The Double Bubble TOE permits the known laws to be extended below the mobic scale in reciprocal solutions that still conserve the laws of physics. Thus Einstein's term of *general covariance* offers a way to generalize the physical laws undergoing reciprocal coordinate transformations between both bubbles. In other words, in this larger view, the Double Bubble TOE describes covariance.

This PPCCL symmetry of the Double Bubble TOE requires serious consideration because the laws of physics in each bubble remain invariant under its combined operations, and the two bubbles are consistent with each other in a reciprocal, covariant, mobic-mirroring way. The laws of physics are conserved between the two bubbles in this TOE. What a pretty PPCCL!

9. Physical reciprocity works

After exploring PPCCL covariance, now we'll consider ⚏ **Question 10: Why do so many standard model equations have "nonsensical" reciprocal solutions that later turn out to reveal something important…a new particle or law?** Physicists have found reciprocal solutions in their equations that seem to make no sense at first. Then someone achieves a scientific breakthrough that suddenly renders one of those "nonsense" equations quite meaningful.

In 1928, Paul Dirac made an equation based on relativity proposing that the electron can have either positive or negative energy as solutions. Many scientists doubted Dirac's nonsense equation, but a hunt started to resolve it.

In 1932, a new particle appeared in Carl David Anderson's cloud chamber that fulfilled the "nonsense" reciprocal solution of Dirac's equation. It was the anti-electron (often called the positron). Using that finding as a model, Dirac predicted the existence of antimatter for other specific particles, and over time, more discoveries validated the generalized condition of antimatter.

That's just one example of the stunning reliability of prediction that comes from examining reciprocal symmetries. Seemingly "nonsensical" reciprocal solutions exist because the universe has innate symmetric reciprocity that started in polarity itself. We live in a Double Bubble universe that began when polarized number bifurcated in binary and analog modes and developed an analinear master code of polarized pulsing in 64 patterns of fractal co-chaos.

Those patterns manifest the four primals in two complementary, holographic bubbles. Dimensionality's double trident of reciprocal space and time spans both bubbles of our living universe like an invisible skeleton, and reciprocal versions of matter and energy ride upon it like flesh.

After the master code used the 4 primals to establish the mirror-twin bubbles in reciprocal formats, it templated many lesser code variants in each bubble. Our bubble has the genetic code, the I Ching math, and many octaves of broken symmetries that are evident in physics, chemistry, and music.

Science, wake up and look closer! So many clues up here suggest that this holographic universe contains a mirror twin hidden below the mobic scale with a reversing-mirror format of timespace and tachyonic antimattergy and the lost pole of gravitation. Recognizing the Double Bubble structure can help us clarify and resolve some major puzzles of current science.

10. Mental reciprocity works

Mirror reciprocity exists between the two bubbles not only in their physical features but also in their mental makeup. Above the mobic scale in this bubble, mind power by electricity evolves in many separate, small, fractal-based organisms existing in constant *now*. Like you and me, for instance.

Meanwhile, below the mobic scale in the other bubble, mind powered by tachyonic energy evolves in a single giant, buzzing cloudbank existing in constant *here*. Its fractal-based patterns are stable in form, yet their contents are paradoxically also in constant flux. They support the overall evolution of the universe and foster the wellbeing of us walkabout organisms.

Aspects of that great mind interact with our own tiny minds. We tap into its larger patterning occasionally in the daytime but more often at night. We perceive its flow in what we call dreams, archetypes, premonitions, mythology, memes, morphogenetic fields, Akashic records, synchronicity, telepathy, remote viewing, collective unconscious, noosphere, divine plan, Tao, and other names, too.

Even as entropy would seem to be the destiny of our universe's physical body, negentropy is the thrust of its evolving consciousness. Our universe's long-term goal is to know and be known in all its parts, to realize Self and Other by developing its capacity to the fullest degree possible, given its parameters.

Is it possible to communicate with that wisdom consciously? Sure. How? Prayer. Meditation. Dreams. Nature. And in other ways, too. As a supremely aware virtuoso in the art of possible futures, the giant mind below the mobic scale can communicate with our diverse minds up above it to further events.

For instance, in our sciences, arts, talents, and hobbies, individuals are often inspired by a fluke of illumination that seems to arrive "out of the blue" or "from nowhere." It happens when we tune into a wavelength of creativity beyond our own capacity. We draw unconsciously on greater mind when "out of the blue" comes a new idea, insight, or invention. We can access that invisible support in what Carl Jung called synchronicity, in what ancient China called the Tao. We can even tap into it with the I Ching or whatever else helps us recognize and respect the greater mind cultivating itself by cultivating us.

11. Our universe has aspirations

Our particular universe is testing the viability of love over the long haul. Love turned creation into more than just a zero-sum game by recognizing its nothing as something, the number 0, able to exist or not, bifurcate, polarize, count and relate, becoming ever more diversified, ever more complexly and completely aware of the grand unity beyond all its diversity.

Even though death appears to be the destiny of our human bodies, negentropy shapes the thrust of our ongoing souls nestled in the tachyonic cloud of the lower bubble, reincarnating again and again via fragile fleshly forms in the upper bubble. A record of each lifetime is kept in the lower bubble, and as those successes get more successful, our soul identities will eventually become part of our conscious lives.

Meanwhile, reincarnation is occurring for the universe, too. Its living structure continuously rebirths itself, emerging from instant to instant at the mobic scale, again and again evolving, motivating its great life toward a goal that philosophy might call teleological, or religion, or the divine plan. Our living universe tries to amplify its successes while completing its current long lifetime.

Then, because its 8-looping time vectors in the upper bubble extend slightly above its 3D space, the universe will retain an imprint of that final reality as it cocoons up briefly into the firm, foam, foursquare bolster mentioned in Chapter 14. Using that last 8-looping imprint as its DNA, the universe will reincarnate into a huge, filmy light body that is ephemeral yet seems nearly eternal, creatively light-hearted and full of fun, rolling in endless play with the divine.

Reincarnation at a universal scale will enact the very paradigm that we tiny humans here in this upper bubble already anticipate as we busily accrue our bits of wisdom gained in life after life, peat and repeat, to stoke the larger fires of awareness. So you do matter…even though this Double Bubble TOE describes but one of God's many TOES upholding all the universes beyond legion.

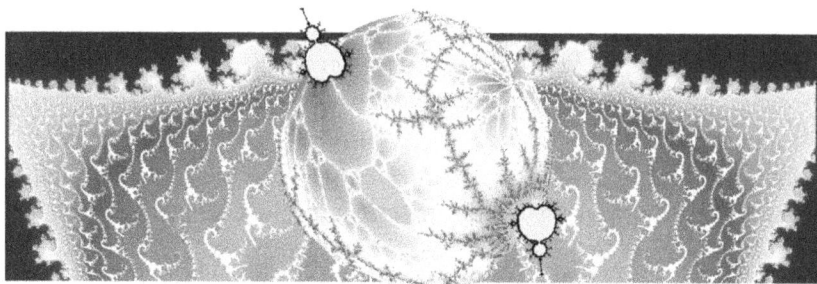

Chapter 18: Hexagram 4

Tapping into the I Ching offers a two-way conversation with the Tao. In each volume of this series, the last chapter explores some aspect of I Ching interpretation by combining both logical sequencing and analog examples.

This Volume 4 examines Hexagram 4, and the interpretation is my own. Below you'll see the hexagram number, its name in Chinese and English, and its hexagram, a mathematical figure. After that, you will find the *Image, Judgment, Hexagram Lines, Line Interpretation, Analogy, Analysis,* and *Example.* First, read the *Image* and *Judgment* to get the basic dynamic of the hexagram.

1. Hexagram 4

Hexagram 4: 蒙 *The Novice* 　 *Co-chaos Math*

The Image

Fog rises at the mountain's foot,

At first, obscures the peak—

Slowly it resolves to dew

That limns in chasm and creek

and perilous path to philosopher's stone,

If you but humbly seek.

The Judgment

I do not seek out the young novice. The novice seeks me.

When the beginner first asks to divine my meaning, I instruct.

But when the same question is asked again and again,

that is tiresome begging.

If the novice importunes without listening to me,

I cease to answer him.

Benefit comes from being resolute and true,

for it betokens progress and success.

IMPORTANT: This hexagram's *Judgment* contains the only time in all 64 hexagrams that the I Ching uses the first person singular "I" as the voice of the I Ching itself. This voice wants to instruct, but it becomes annoyed if the seeker asks for an answer, then refuses to accept it, and keeps trying for a different, more pleasing answer. Lines 3 and 4 mention other ways that a troublesome pupil can also go wrong by refusing to learn after instruction on a topic.

The Lines

Hexagram 4

MOUNTAIN

WATER

Line 6
Line 5 ☆
Line 4
Line 3
Line 2 ☆
Line 1

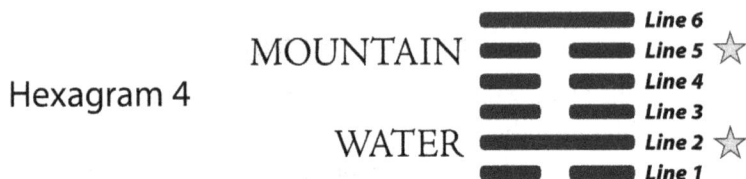

☆ **denotes the most important line(s)**—*usually 2 and/or 5.*

Now overview all the hexagram's numbered lines to understand its dynamic. **Line 1** sits at the bottom of the hexagram figure. Read upward to **Line 6** at the top. Next, apply any changing lines *in your own answer* to add the nuances of their dynamics *to your own specific situation*. Add to that any insights you may find in the *Image, Judgment, Interpretation, Analogy, Analysis,* and *Example.*

Line 1 ▬ ▬
Rigorous discipline breaks mental shackles,

But ceaseless correction raises the hackles.

Line 2 ▬▬ ☆
A wise mind respects women, accepts naiveté

In a child, and points the way.

Line 3 ▬ ▬
Attracted to yang in greedy obsession,

Yin foolishly loses her own self-possession.

Line 4 ▬ ▬
Encircling walls of blind ignorance

Forge chains of passive impotence.

Line 5 ▬ ▬ ☆
Eager to learn, open-minded, unafraid,

Such childlike thirst attracts the right aid.

Line 6 ▬▬
Why flog ignorance, one's own or another's?

Gain comes from warding life's blows off our brothers.

2. The Line Interpretation

Next, ponder the hexagram answer. Consider its general tone, modified by the specific influence of each changing line you received in your answer… it's like adding varied sprinkles atop your ice cream cone's basic flavor.

Hexagram 4 is about the difficulties and responsibilities faced as a novice in life when one is young, or when mature, even as a novice in a new situation.

Line 1 __ __

At the beginning, before muscles are hardened and lungs are deepened to hold more oxygen, the student will feel punished by the grueling climb toward knowledge. Planning, discipline, and guidance are needed for the learner to make progress. But too much discipline is equally discouraging. Thus when a foundation is acquired, the student should be trusted with increasing freedom.

Line 2 ____ ☆

According to the early Zhou, the enlightened male head of the family should become a model who teaches tolerance to others. He should be broad-minded about the naiveté that causes mistakes and model how to learn well. He should not be angered by his wife, nor threatened by the vigor of his son, for each member can help sustain the whole family. It suggests that one in a position of power should act with forbearance to those who seem ignorant, weaker, or younger, for their support becomes precious finally. This powerful Line 2 yang, as good teacher, is the natural companion of Line 5 yin, embodying the good student.

Line 3 __ __

This yin line portrays a wobbly indecision teetering between two objects of desire. Yin's inaction is due to wavering on the cusp between choices. Yin spies powerful yang up at the top of the hexagram and fancies such a valuable-seeming partner. But the yang line just below seems closer and more familiar, so yin splits her desire between the near and the faraway attractions. Inconsistent and vacillating, yin loses her own identity by indecision between the two attractions.

The text says literally, "Do not marry [the] girl. Seeing golden man, no longer possesses [her own] body." It means do not "marry" passivity in this situation. In the inability to choose between alternatives, a person loses oneself. Shilly-shallying between two seemingly-glittering routes hinders moving up the slope of life.

Line 4 __ __

As the slope of life's learning rises higher, so does the risk of loneliness, even of getting lost on a new or untrod path. One feels hemmed in by a blind canyon with steep walls of ignorance that offer no way out. Due to the lack of solid footing in a fruitful direction, this is a discouraging time, but this bottom yin line in the upper trigram of Mountain counsels to hold onto your aim and be patient.

Line 5 __ __ ☆

This yin line is in the most-favored position in the hexagram. It indicates youthful receptivity that asks to be taught. Yin's qualities of gentleness, humbleness, devotion, and receptivity in this favored fifth position suggests by analogy your own willingness to be taught. A sincere seeking will meet with good fortune. Its proper teacher is yang line 2, boding well, for they make a good match.

Line 6 ———

This final yang line shows precarious strength above Line 5's yin student. Metaphorical blows to that student could come from life's hard lessons or even a hyperbolic self-criticism from within. But striking the neophyte brings no gain, so the strength above will not strike in a high and mighty fashion. Rather, it will deflect blows from the novice below; and likewise, the student should cease self-disparagement. Improvement in learning will come by warding off injury.

3. The Associations

Youthful Ignorance, Climbing toward Clarity, Innocent Seeking, The School of Hard Knocks, Immature Effort, Learning Life's Lesson, Trying to Climb Higher, Your own association with this archetype

4. The Analogy

The fractal pattern of Hexagram 4 ☷☵ *The Novice* symbolizes water's remarkable power to go up as well as down. Its trigram of *Water* ☵ starts out as a canyon fog that obscures the lower slopes of the mountain when the young learner sets off to climb life's rocky path. Through the progressive development of the lines, that water clarifies and resolves into dew that outlines the path. Eventually, the water as vapor ascends the *Mountain* ☶ to become clouds, before it falls again as rain and flows down the mountain again as rivulets turning into creeks and canyons. This hydrologic cycle of water allows life to prosper.

The dynamic here is the natural next step after birth: the uphill lessons about life. One climbs the mountain of humanity's codified knowledge and material culture toward the lofty heights of spiritual wisdom. The huge heritage of the past is a silent teacher that helps the novice move upward toward wisdom.

In the *Judgment* of this hexagram, the teaching voice of the I Ching itself speaks to the learner, pointing out, "I didn't seek you; you sought me." When the I Ching is asked a question, it will answer to teach the learner.

But if that answer is refused or ignored, the universal wisdom realizes it and lets the learner realize it, too. Life's lessons should not be ignored; looping into the same mistake again and again indicates a heedless or poor learner. Sage wisdom lives high on the mountain peak, but the climb to reach it is hard.

5. The Analysis

Hexagram 4 *The Novice* is a dynamic that comes quite naturally in progression after Hexagram 1 *Assertive Heaven*, Hexagram 2 *Receptive Earth*, and Hexagram 3 *Laboring Birth*. As soon as the amniotic sac breaks and releases its watery flood, the infant's education begins. (Note that the yin-yang line structure of Hexagram 4 inverts that of Hexagram 3.)

The lower trigram of *Water* ☵ acts at first like canyon fog obscuring the upper trigram of *Mountain* ☶ , which represents the huge pile of life experiences that one must climb, clambering up by means of knowledge into the peaks of understanding. The novice starts up the mountain path, only just beginning to realize how much learning looms ahead. Sixty more hexagrams await exploration as the young student sets out on the hard task of discovering the world beyond the dynamics of Father, Mother, and Birth.

At the start of the climb up the mountain, its looming but undefined mass appears huge and obscure ahead. It is a *massa confusa,* yet it leads the way toward life wisdom. That goal is blurry, out of focus in the mist of the lower levels, vaguely threatening in its glimpses of hidden precipices, ravines, rocky slopes, but there are also hints of glorious summits of vision—yet all of it is still shrouded by the blurry fog of ignorance.

The lower trigram of *Water* denotes the beginner's path that makes a hazardous journey through canyon walls on either side. Line 1 suggests that discipline will sharpen the learner's focus and so begin to clarify the foggy landscape—yet too much rigid discipline is as thwarting as too little.

In Line 2, the student discovers that tolerance can turn irritating obstacles into assets. As one slowly climbs, the resolving dew will scintillate on the terrain of events to highlight a craggy beauty that reveals the way toward wisdom. But the path up the mountain in search of the Philosopher's Stone is steep, and it offers many wrong turns, cul-de-sacs, and false trails toward fool's gold.

In Line 3, unwilling to settle on a nearby choice, the student notices the distant glitter of something that promises gain by association. Torn between options and blinded by split desires, the student loses the way to self-realization.

In the upper trigram of *Mountain,* one sees the mountain, not the fog. At first, however, its mammoth bulk of rock also is daunting.

In Line 4, the learner feels completely hemmed in by rocky walls and vertiginous space, with no alternative remaining but to hold on patiently.

Line 5 achieves the right combination of eagerness and humility that will let learning come easily. It signifies that the most important factor in surmounting life's lessons is an attitude of humbly seeking the truth.

Line 6 at the top suggests that yang power up here should not be used arrogantly to strike down learners while preening oneself with an overbearing satisfaction. No advantage comes from lording it over the untutored; rather, gains come from fending blows off the learners so that they can learn more easily.

The dynamic of this hexagram describes the learner-teacher dyad and appropriate behavior for both positions. Sometimes in life, one is the learner, and sometimes one is the teacher. It is even possible to be both simultaneously, for the good teacher will learn even while teaching, just as the good student is willing to share acquired knowledge. An honest teacher honors truth over prestige and is willing to admit mistakes and learn from them, thus becoming a better teacher.

Some translations have made this hexagram seem unduly censorious toward the ignorant, calling the young student a fool. But a novice is not innately foolish. Learning is essential to achieve a more abundant life, and learning is endless. Rather, ignorance is foolish only if it is arrogant, not bothering to do reality checks on its assumed knowledge and make frequent self-corrections.

In summary, perseverance, humility, and a good guide who can help ward off injury—these make it easier for the student to ascend the path of learning to a broader vision. The massive mountain of accumulated knowledge, wisdom, and understanding may defeat the climber, but wise pacing can transmute that rocky path slowly into the concrete stability of the Philosopher's Stone.

Be disciplined, be tolerant, and do not become lured by false glitter. During an impasse, sit tight without losing sight of the goal; keeping a humble eagerness to learn will eventually bring benefits and win protection from above.

6. The Example

Back in 1985, I told my M.D. psychiatrist-turned-Jungian analyst, Dr. Joseph Wakefield, that I was starting to learn about the I Ching. He responded by saying that he'd fooled around with it himself during his pre-med training at Stanford. As he spoke, his intonation hit hard on the word *fooled.*

"So is it foolish to use the I Ching? A waster of time?" I was baiting him to see what happened. "You think it's superstition-ridden claptrap?"

"Oh, no," he replied. "On the contrary. But I certainly believed it was foolish at first."

Then he told me that he'd mocked some Palo Alto friends who were using the I Ching. He finally annoyed them so much that they suggested he try it out instead of just passing blind judgment. Now I'll try to recreate our conversation from my notes and memory. It went something like this.

Wakefield said he asked the oracle a jeering question: "I'd be a fool to ask you a question, wouldn't I?" Wakefield looked at me wryly. "Then I read the answer. It scared me so much that I dropped the I Ching book like it burned my hands. I stayed away from the I Ching for 10 years."

"What answer did you get?"

"Hexagram 4—*The Young Fool.*"

I laughed. The Chinese name of Hexagram 4 is also translated as *The Young Fool, Youthful Folly, Inexperience,* and other such phrases. I call it *The*

Novice because this hexagram symbolizes the dynamic of a youthful novice, an innocent, or maybe an older ignoramus who does not realize what's going on.

Scholars have long emphasized that this is the only hexagram where the I Ching speaks in the first-person voice of the Tao itself. Here is James Legge's translation of Hexagram 4: "I do not (go and) seek the youthful and inexperienced, but he comes and seeks me. When he shows (the sincerity that marks) the first recourse to divination, I instruct him. If he apply a second and third time, that is troublesome; and I do not instruct the troublesome. There will be advantage in being firm and correct."

When Wakefield told me about his introduction to the I Ching, I already knew that a hexagram's message is explained by a few terse sentences using long-ago analogies from a rural, agrarian culture in feudal China. I'd also begun to surmise that the yin and yang figures might be a mathematical shorthand for the hexagram's dynamic.

Wakefield went on, "I read the text, and while I was reading about *The Young Fool*, I felt like one…my face was burning, a young fool chastised. I was also afraid of what might be pointing that out to me. So I called it superstition…" he laughed, "…which is a different thing, I found out years later, from what actually occurs with the I Ching."

"What do you mean?" I was trying to get his idea of what really occurs with it.

"Superstition is rigid little rituals frozen around symbols. Black cats are evil, so you don't let one walk in front of you. A four-leaf clover is lucky, so you pick it. Spilling salt is unlucky, so you throw some grains over your left shoulder. Blue wards off the evil eye, so you put blue on the baby. That sort of thing.

"Those rigid little superstitions will fossilize your significant moments and turn them lifeless. It will praise a bird song even as it kills the singer, stuffs it, sets it in a niche, worships it as the symbol of joy while playing that same old bird tape forevermore. Then he intoned in a goofy voice, "The stuffed bluebird of happiness…."

I laughed. "And the I Ching doesn't do that?"

"No. It is an image provoker for your mind to dialog with. You consider its dynamic in your own unique situation. So the symbol itself stays alive as its pattern flows and shifts in its details, depending on what you need to intuit at a given time. The process is subjective, and your I Ching answer actually relates to you. It shows whatever you need to see now. Whether you like it or not."

"But how? Why?" I asked. By this time, I already knew from experience that the I Ching appears to work…somehow!…so I was testing my secret surmises and conclusions against what Joe Wakefield was saying.

I still did not feel ready yet to declare my own experiences with it openly. Not yet. After all, I taught at the University of Texas. Talking freely about the I Ching might invalidate my professional status card.

Or hmm…maybe I just needed to discover a different crowd?

"I don't know," he said. "Beats me. Synchronicity, Jung called it. An acausal connecting principle, he said, and Wolfgang Pauli agreed. Whatever it is, it works. Oh yes, even in that archaic language, the I Ching's allusive wisdom is beyond chance or superstition. Jung realized it. And I found it out eventually." He smiled. "I'm no longer such a young fool. After 10 years I came back to the I Ching. Now I use it often."

Interesting, isn't it, that it took 10 years for the message to get past his conscious resistance? Sometimes it takes a long time because the I Ching is so gentle, so silent, and so abstract. It does not flay you, excommunicate you, or shun you. It is not wreathed in lightning, neon, or even flesh.

I have slowly learned that the I Ching reveals the patterns of events. Not their specifics, but the underlying dynamics. It operates via fractal co-chaos that can describe or predict a trend without specifying its exact details. Finding this huge intelligence that resides deep in the weave of nature, even learning to communicate with it, can be disconcerting, even frightening…until it becomes wonderful.

7. Adele Aldridge: Hexagram 4, Line 2

All of the books in this series conclude with an image and commentary about a particular hexagram. The mathematical hexagram figure and some verbal metaphors are used to convey its co-chaos pattern. Because this is the fourth book in the series, this chapter deals with Hexagram 4, *The Novice*.

The dynamic of Hexagram 4 is likened to the novice, youth, student, ignoramus, or fool standing at the bottom of the mountain of knowledge, wondering how high to climb, whether it is worthwhile to do so, or if it is even possible to do so. Yes, the climb is both worthwhile and possible, but it takes fortitude, patience, tolerance, and a desire to do the right thing.

Now I'll show you how Adele Aldridge interprets this hexagram…well, at least one line of it. She is a longtime explorer of the I Ching in art and meditation. Adele and I met online years ago through our shared interest in the I Ching. We both realized it speaks to each person differently, uniquely, even while it holds the whole world in its grasp. We both know that when you get an I Ching answer, it shows the dynamic that is iterating its unique variation on a co-chaos pattern operating in your life. Once you recognize the pattern, your solution begins to emerge…if you are looking for it.

Hexagram 4
THE
NOVICE

In the following image, Adele illustrates Hexagram 4, *The Novice,* Line 2:

Hexagram 4, Line 2

-:::-

With the strength of youth and little knowledge
I am tolerant to what is hidden and undeveloped.
My true power unfolds into a larger and freer self
After I absorb and transform experience into character.

-:::-

I chose to show you Line 2 of Hexagram 4 because it portrays how ignorant I felt when I started working on this series in 1985. I read physics, math, and biochemistry, hoping to recognize some formal concepts and terms that would explain what that great dream showed me. I stumbled my way up the mountain of learning, falling over boulders and into crevices again and again. By filling in an information gap, I'd pick myself up and go on.

Mining the riches of that dream took more patience and deep meditation on passing events than I was used to. For a long time, I felt trapped in the cocoon depicted by Adele's image for Hexagram 4, line 2. Without knowing how to spin silk, I found myself trapped in a cocoon of ignorance. I didn't know how to emerge into a winged state, or if I did, what to do with all that lovely silk.

But I experienced the cocoon's ripening effect. Slowly I metamorphosed into understanding at least enough to explain some aspects. What a dazzling, bewildering, challenging, fascinating, ongoing journey this has been!

8. Probe the I Ching sources

The I Ching is one of the oldest books in literature. When I was learning to use it, I often opened four or five different books, comparing one text to another like various friends discussing the hexagram from their slightly different viewpoints, seeking what resonated most in it all for me and my particular question. That's how I learned to understand the I Ching's analogies.

The internet is a modern treasure trove of information. Hilary Barrett's I Ching study site at underlineonlineclarity.co.uk lets you improve your understanding of the I Ching and share questions. You can also purchase her book *I Ching*, which I recommend. It contains a succinct I Ching translation of each hexagram. It handles I Ching imagery, history, and key concepts well. However, rather than employing the coin method described in that book, I suggest that you instead use the 16 stones system for consulting the oracle itself.

You can find free I Ching programs and thousands of years of relevant texts on the web. The ebooks in this series provide links to many sites. There's a wide diversity of approaches to the I Ching. For example, Roger Sessions, Benedictine Oblate, wrote the book *Wisdom's Way: The Christian I Ching*.

The I Ching is so old and encompassing that it can embrace many approaches from different ages and cultures. I invite you to explore its historical depth and inclusive breadth without end.

Series Summary

1. What is our universe?

This TOE says we live in the Double Bubble universe. Its two bubbles are conjoined, symbiotic mirror-twins with reciprocal properties of space, time, matter, and energy. Science sees our white-hole bubble above the *quantum* scale, where matter and energy emerge. It does not see a black-hole bubble conjoining our bubble at the far-tinier *mobic* scale where space and time emerge.

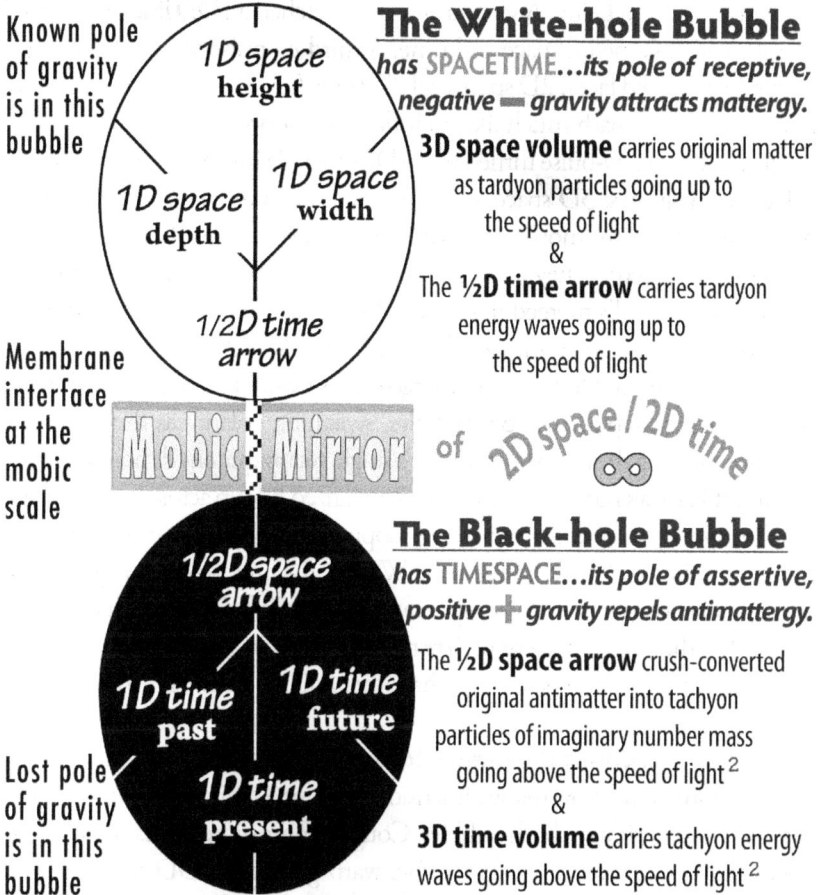

Known pole of gravity is in this bubble

1D space height

1D space depth

1D space width

1/2D time arrow

Membrane interface at the mobic scale

Mobic Mirror *of* 2D space / 2D time ∞

The White-hole Bubble

has SPACETIME...*its pole of receptive, negative ▬ gravity attracts mattergy.*

3D space volume carries original matter as tardyon particles going up to the speed of light
&
The **½D time arrow** carries tardyon energy waves going up to the speed of light

The Black-hole Bubble

has TIMESPACE...*its pole of assertive, positive ✛ gravity repels antimattergy.*

1/2D space arrow

1D time past

1D time future

1D time present

Lost pole of gravity is in this bubble

The **½D space arrow** crush-converted original antimatter into tachyon particles of imaginary number mass going above the speed of light ^{2}
&
3D time volume carries tachyon energy waves going above the speed of light ^{2}

The Double Bubble universe has 11 dimensions

Our upper bubble has the *spacetime trident* of contiguous 3D space with a one-way, ½D arrow of time, plus one pole of gravitation and the original matter. Its 3D space holds many material 3D structures morphing on the arrow of time, but its particle-waves are slowed down to the speed of light.

The lower bubble has the *timespace trident* of 3D time with a one-way, ½D arrow of space, plus gravitation's "lost" pole, plus the original "lost" antimatter that was long ago crush-converted by that lower bubble's meager ½D space into tachyon particle-waves going above the speed of light[2]. It powers a huge, unified mind constellated in vast 3D energy patterns in the lower bubble's 3D time.

The mirror-twin bubbles conjoin at a membrane interface of ultra-tiny, mobic pores called *mactors*. Each pore combines traits of a Mobius band and a Lorenz attractor, hence the name of *mactor* for its dynamic at that ultra-tiny scale.

How did space and time begin? The cosmegg set a dimensionless point with an *on*-pulse of being. A second *on*-pulse sketched a 1DD line of polarized tension with two poles: space and time. A third *on*-pulse set a 2DD triangle with two polarized faces: 2D space and 2D time. Polarized tension ran around both faces on a X path much like an infinity ∞-loop.

Just one more *on*-pulse turned that 2DD-triangle into a 3DD tetrahedron. It had two volumes: 3D space and 3D time. The outer volume of 3D space projected far above the mobic scale, while the inner volume of 3D time projected far below it. Together they made an hourglass cell. That single cell replicated many times. All cells merged into our holographic bubble of 3D space above the mobic scale; below it, into a holographic bubble of 3D time.

The Double Bubble hologram merged all its ∞-loops and projected them, so now they 8-loop across both bubbles, switching polarity as they cross the interface, creating the tensor network of a single yet ubiquitous dimension made of two ½D arrows moving on an endless, polarized 8-loop across both bubbles.

The upper and lower halves of this 8-looping tensor network are polarized per bubble as either ½D time or ½D space. We have the time pole. We experience it as the point of constant *now* moving forward on the arrow of time. But that other bubble has the constant point of *here* moving backward on its arrow of space. *Here* is the only location possible in that other bubble. (There is no *there* there.)

2. Count the dimensions of this kleiniverse

Our universe has how many dimensions? Count 3.5 dimensions per bubble, making 7 dimensions in both bubbles. Count 4 more dimensions at the mobic scale itself, where every mactor's mobic warp generates 2DD triangles with polarized faces of 2D space and 2D time. This totals 11 dimensions in a layout of complementary space and time that is symmetrical across both bubbles.

The dynamic of the Double Bubble recalls a Lorenz attractor. Its two domains are the upper and lower bubbles. Its 3D space and 3D time act as three coupled Ordinary Differential Equations (ODEs) iterating along the arrows of ½D time and ½D space to evolve the nonlinear solution of reality emerging in both bubbles. The reciprocal laws of physics and the reciprocal scaling of space and time, matter and energy let both bubbles fit inside each other as a *kleiniverse*.

3. The master code uses four primals

This TOE says our universe is a huge, living organism whose fractal structure is generated by a co-chaos paradigm that iterates in self-similar patterns on many scales. Its master code uses four primals: space, time, matter, and energy. This polarized pair of pairs sort into two *carriers*: space and time…and two *cargoes*: matter and energy, polarized such that space carries matter, and time carries energy.

↓ Carrier Pair	4 Primals	↓ Cargo Pair
1. Space	←⋯ *CARRIES* ⋯→	**3. Matter**
2. Time	←⋯ *CARRIES* ⋯→	**4. Energy**

Panel: the 4 primals are a polarized pair of pairs

Our 3D space bubble has tardyonic particle-waves in vast material structures of self-similar, evolving 3D patterns. The 3D time bubble has speedy tachyonic particle-waves in vast energetic constellations of self-similar, evolving 3D patterns. In both bubbles, intricate detailing on many scales recalls the Mandelbrot set.

Both bubbles cooperate to refresh their space-time forms and update their matter-energy cargoes at a rate that makes our holographic universe appear to be smoothly continuous to our senses and mechanical tools above the quantum scale, the smallest scale known to current physics. In the universal body, old configurations decay and new ones develop. We tiny organisms in the upper bubble experience this flux as the emergent events of ongoing reality.

4. The genetic code is a variant of the master code

How can we decipher the master code that iterates the universe? We can study a lesser variant, the familiar genetic code that iterates us. It offer us some clues.

CLUE: DNA uses four base molecules: **T**hymine, **C**ytosine, **A**denine, and **G**uanine. Its polarized pair of pairs sort into two *pyrimidines*: **T** and **C**—and two *purines*: **A** and **G**. They are polarized such that **T** bonds with **A**, and **C** bonds with **G**.

CLUE: The four base molecules can pair-bond by triplets (*codons*) to make 8 × 8 = 64 molecular 6-packs on the double helix to iterate and maintain the bodies of all evolving species. Old configurations of individual organisms decay, and new ones develop. We experience this flux as the emergent lives of ongoing species.

↓ Pyrimidine Pair	4 Molecules	↓ Purine Pair
1. T	←⋯ *BONDS WITH* ⋯→	**3. A**
2. C	←⋯ *BONDS WITH* ⋯→	**4. G**

Panel: the 4 DNA molecules are a polarized pair of pairs

5. Our Rosetta Stone: I Ching, genetic code, master code!

This TOE says the ancient I Ching hexagrams of China offer a math shorthand for this paradigm. It grows on a bifurcation tree that is both doubled and polarized, i.e., it is a *dp-tree*. The I Ching's easy math can shorthand the genetic code, and their kinship gives us a Rosetta Stone with three scripts, two known and one unknown. The two known codes can help us decipher the unknown master code.

CLUE: Like DNA, I Ching math figures are also a polarized pair of pairs. They sort into two stable bigrams and two unstable bigrams: stable yin ☷, stable yang ☰, changing yin ⚎, and changing yang ⚏.

Yin-based ↓   *4 Bigrams*   **Yang-based** ↓

stable yin   **1.** ☷   ←⋯ *STABLE PAIR* ⋯→   stable yang   **3.** ☰

changing yin **2.** ⚎   ←⋯ *CHANGING PAIR* ⋯→   changing yang **4.** ⚏

Panel: the 4 bigrams are a polarized pair of pairs

CLUE: The I Ching math develops on a dp-tree, and it can shorthand the other two code variants. Below, the dp-tree has three levels of polarized forking above and below a neutral 0 seed in the middle. On the dp-tree, *minus* − stands for a yin ━ ━ fork. *Plus* + stands for a yang ━━ fork. Its first level of forking outward develops −*yin* and +*yang* poles. The second level outward develops the four *bigrams*. The third level outward develops eight *trigrams* above and below.

Each trigram is a *vertical* period 3 window (*vp3*) defining a chaos pattern. This postulates a vital variant on Yorke and Li's *horizontal* period 3 window (*hp3*) that defines a chaos pattern in their seminal paper *Period Three Implies Chaos* (1975). In each vp3, addition by 2s, period-doubling, and exponential power together create a nonlinear chaos process so special that I call it *analinear*.

The dp-tree has 8 × 8 polarized vp3s = 64 co-chaos patterns

CLUE: Each trigram's math describes a *chaos pattern*. The dp-tree can pair-bond its trigrams into 8 × 8 = 64 *hexagram* 6-packs of *co-chaos patterns*.

Our Rosetta Stone's triple play features the familiar genetic code, the ancient I Ching, and an unknown master code. The first two codes have some shared traits that will help us discern features of the master code at the mobic scale, where polarized pulses organize into triplets of information in myriad mactors. The triplets then pair-bond into 8 × 8 = 64 co-chaos patterns that develop our universe's emergent properties. They project the Double Bubble's space-time skeleton and flesh out its matter-energy body to evolve its huge, ongoing life.

6. What are we?

This TOE says our Double Bubble universe lives, and we are like microbes living in its gut, oblivious to its larger aims. In our white-hole bubble of 3D space, we tiny, diverse, walkabout minds are powered by particle-waves of slow tardyon energy. But the black-hole bubble of 3D time holds a single, giant, unified mind that is powered by zippy tachyon energy moving at more than lightspeed[2].

Many tiny bodies with portable minds inhabit the upper bubble. But when a mind is released from ego identification by sleep, trance, meditation, or other means, it can tap into aspects of that unified mind in the tachyonic cloudbank of 3D time (some call it God or Mother Nature) processing the data of concurrent past, present, and future. That unified mind evolves its huge, beautiful, and diversified universal body. For instance, in the upper bubble, it established the far-flung galaxies and tiny micro-organisms under rocks that hold our attention.

The universal mind even delivers dreams to us nightly, in dramas that address our specific needs, fears, and hopes…but most of us have forgotten how to translate its symbolic lingo. Relearning it can cultivate a sixth sense, an ability to tap into nature's basic patterns by deep-see diving. We can access info in the tachyonic cloud of the lower bubble via shared intention and resonance. It recognizes and responds to whatever is in you—so go carefully and with good intentions. Treated wisely, it can heal and unify us, body and soul, layer by layer.

The minds in both bubbles contribute in various ways to the thrust of universal existence, which has a greater purpose beyond our own human preoccupations. Our universe plans a wider future for us as we become more conscious of our place in the whole. It has been patiently cultivating its universal life, including us among its myriad forms, hoping to evolve us enough to recognize that it too lives…and further, coaxing us to divine that there is something even greater beyond. We have the chance to acknowledge, share, and improve this destiny.

Blurb and Reviews

The Universe Is Alive and Well
The Organism

Katya Walter, PhD
Biography

Katya Walter has a Ph.D. with an interdisciplinary emphasis from the University of Texas at Austin. She spent 5 years of post-doctoral study at the Jung Institute of Zurich, and a year of post-doctoral study in China. Dr. Walter taught in colleges and universities in the USA and abroad for 16 years before focusing on writing and lecturing. She has given numerous workshops on the I Ching, chaos theory, synchronicity, and dreams in the United States and Europe.

-:§:-

From the Editor

This book is Volume 4 in the dazzling *Touching God's TOE* series, 4th edition. In this volume, Dr. Walter discusses our universe as a living system. She describes the master code that created its physical structure and the universal mind that vitalizes it. While offering scientific concepts, she also discusses philosophical parallels and dreams that inspired her to write this series.

This book has 18 chapters in 115 sections, with 79 listed images, graphics, and charts. It includes a *Series Summary*, *Bibliography*, and *Reviews*. The more science-based chapters have odd numbers (1, 3, 5, etc.). The more philosophical chapters have even numbers (2, 4, 6...). The ebook version has an interactive table of contents and 125 e-links that act as informative footnotes. Its text is searchable and receives electronic updates. It is hand-edited to hold color graphics that allow greater distinctions in images and charts. Consider getting both the print and ebook versions for a greater range of information and versatility.

Science and mysticism merge in this stunning new paradigm. China's ancient I Ching followed the flow of the Tao, universal mind. Western science investigates the genetic code that generates organic matter. This paradigm merges them to explain both mind and matter. Let Dr. Walter guide you through the patterns of chaos theory into mystic beauty! This book heals the 2,500-year split between body and mind. It embraces East and West to unite our planet. Called in Germany a *"philosopher queen of the global village"*... she *"merges left-brain accuracy with right-brain vision! Scientific truth speaks in a clear voice of wisdom."*- **Claus Claussen**

-:§:-

PRAISE FOR THE *TOUCHING GOD'S TOE SERIES*

What an interesting and inspiring writer...interesting scientifically and inspiring metaphysically! I have traveled widely, but never on a roller coaster of dimensionality before! It makes Flatland look—well, flat. Quantum organics reveals how space, time, matter, and energy mirror aspects of our DNA. And the author's take on what animals think is shockingly possible! You'll never regret picking up this series and reading it. It will take your mind to new places, and it will lift your soul along the way.

Lynn Hayden
Consultant, Singapore Institute of Management

"I find the *Double Bubble Universe* the most promising of all the TOEs being proposed currently. It involves a new topological model spanning all levels of reality and "deep-see diving" into fractal pattern recognition. It answers far-reaching questions such as 'How did our universe begin?' and 'How are telepathy and remote viewing possible?' This model deserves careful reading by the best minds of our time."

Oliver Markley, Ph.D.
Professor Emeritus, Human Sciences & Future Studies

"Are you smarter than a fifth grader?" Better yet, can you bring the clarity of a child's fresh perspective to a Theory of Everything (TOE) that reinterprets standard physics data to reveal a stunningly new and elegantly symmetrical model? If you can, then this book's for you.

Dr. Katya Walter shakes the foundations of currently accepted concepts about physics, metaphysics, and the nature of consciousness. She offers a comprehensive exploration of the way fractal chaos theory forms the underlying structural dynamic that creates and allows for the ongoing evolution of both mind and matter. She also addresses the relationship of mind and matter - physically, spiritually, and philosophically - in ways not previously presented elsewhere.

This seems like a good place to mention that no weeping, wailing and gnashing of teeth are required when reading this book...it is well written in a way that is comprehensible to a general audience as well as for scientists.

If you can set aside any skepticism and/or preconceived notions long enough to allow lucid consideration of the concepts she proffers here, you may be the first on your block to recognize her Theory of Everything as the dawn of the brightest new paradigm since Newton.

Brenda Kennedy
Reader

I can't decide if this is fact, sci-fi, or psi-phy. Whatever, it is truly fascinating. A brain gym of possibilities!

Frank Patterson
Aerospace Engineer

My guess is that your natural reader would be a non-scientist who wants to put science and philosophy together in a coherent mental image.

David Booth, Ph.D.
Mathematician, Inventor
-:::-

I cannot emphasize enough how much I love this book. It makes the most current information about quantum physics into a conversation that can span the thinking styles of both scientists and spiritists. Katya is a dedicated dreamer, and a receiver of concrete knowledge in frontier quantum physics. There should be no separation of physics and metaphysics. There should be fluency and grace and relation to both subjects. This book achieves an understandable explanation of our human experience of dimensionality, and of our fractal nature. She proposes a new Theory of Everything (T.O.E.) If you want the most original elegant synopsis of our existence, which uncovers the mysterious forces of nature (including gravity) and therefore our consciousness, buy this book. You will have an "Aha" moment, and then you will be with me, saying, "Every life-student should be so lucky to have been exposed to Katya Walter. Reading *Double Bubble Universe* is like being in Einstein's living room."

Jennifer J. Colbert
Reader

-:::-

Simply brilliant! And I mean that literally. The clearest explanations are the least complex, and Dr. Walter has managed to take ideas from advanced physics, express them simply, then turn around and analyze the physics to present a clear, simple, and straightforward new paradigm for how the universe works. This is the simplest physics book I've ever read, because of Walter's brilliant use of language that makes these complex concepts entirely understandable. The interweaving of her 'journey-to-the-aha' adds a profound metaphysical understanding of how our universe works from the inside out. You won't regret buying this book.

Anne Beversdorf
Reader

-:::-

The author of this extraordinary book has a rare combination of qualities: an astonishing depth of vision and a genuine modesty. *Double Bubble Universe*... is exploring Katya Walter's theory of everything (TOE). A TOE is the Holy Grail of modern physics. A theory that reconciles the billiard-ball predictability of Newton's Laws with the mysterious goings-on at the Quantum level.

Dr. Walter's book proposes that Physics is blind to another domain in the universe which she describes clearly and patiently with easy-to-grasp imagery... the book really gets you thinking. In a book of this scope, it's very refreshing to find that the author has a gentle, conversational style and an open-minded approach towards the reader.

For example, she writes "Consider this a journey into possibility. I don't mind if you treat this as science fiction, science fact, an amusing tale ... or purely just diverting balderdash ... take it as you will and let it take you where it will."

... where it leads is to the I Ching, the ancient Chinese oracle that, according to Katya Walter, has: "unique fractal shorthand in a coded way that can merge physics and metaphysics."

Extensively referenced and full of diagrams—I really enjoyed this book and I'm looking forward to the next one in the series.

Mick Frankel
I Ching consultant-London, UK

-:::-

"This is the best book on this topic I've read and I've read a lot of them. Solid research on the science end without claiming unproven conclusions. The author simply explains her own TOE which she presents in a logical easy to understand manner. I appreciate her ability to speak to both the spiritual and scientific audience. Very thought provoking.

Winnie Hiller
Reader

-:::-

I think Katya Walter is a genius in that she can translate her right brain insights into left brain analysis with striking correlations and patient explanations. In this book, she's drawing on all her others to outline a sort of unifying theory of everything. Her discovery of the primordial pattern embedded in every level of creation, the "Master Code," is as significant as it sounds. Drawing on her first book, "The Tao of Chaos," she explains that this fractal pattern is originally created by the two primal pairs of opposites: space and time; matter and energy. She then follows the natural implications of that pattern to assert that there is a mirror opposite universe to ours of one-half dimensional space and three dimensional time. In her theory the missing or hidden parts of the pattern that we observe in physics (for example, the "arrow of time") are found in that mirror universe that she calls the Double Bubble universe."

Yeah, the "theory of everything" is a big assertion. Katya Walter's ideas are brave and bold- and impossible to prove. But, as a metaphysicist, she can't wait for the astro and nuclear physicists to catch up. Her books are sort of a field guide to physical reality for modern-day mystics. She explains her ideas through the models of biochemistry, a little math, geometry, and what she calls "the shorthand of the I Ching." She also includes her personal thoughts and dreams with her careful explanations of mathematics and physics. I'll admit, the mix takes getting used to. Yes, it's weird- but worth it!

Like any genius, the author is unconventional and eccentric, which could cause some people not to take her seriously. That would be a mistake, as a careful reading reveals an extremely intelligent and logical woman who asks the right questions. She simply doesn't stop asking, and may go a lot farther than most people are comfortable with, given that we may never have scientific proof for any of this. But, in this era of string theory which proposes many additional unknown dimensions, I wish the physicists would read her books. She could point them in the right direction, and may even save them some time with her simple and elegant theory of everything.

Erin Rose
Yoga Teacher & holistic Health Therapist

-:::-

Dr. Katya Walter's book *Double Bubble Universe* unites cutting-edge scientific research with her own inner 'deep see diving'. She accomplishes the incredible feat of inciting a paradigm-shift in the reader (to the realization that a love-intelligence underlies and pervades the physical universe), bringing her TOE to life! (unlike any other TOE I've read). Quantum mechanics, a physics of cold dead space, births 'quantum organics,' a science of the fractal aliveness of the universe!

Dr. Walter skillfully creates an enjoyable, light read—provocative, funny, and digestible, dealing with perhaps the 'heaviest' topic of all—the structure and meaning of the creation and evolution of the universe. Read this book to witness the wedding of science and heart—watch how every whirling particle spins in the same wind as love's art! Who knows what could bubble up?!?

Peter Craig
Licensed Professional Counselor
-:≡:-

I highly recommend this book for those who believe the current scientific paradigm is incomplete, and are looking for new explanations to fill the gaps. The *Double Bubble Universe* is a space-time, matter-energy, symmetry explanation of the physical universe. It introduces a scientific explanation of the physical laws of the universe with 20 questions. Written in layman's terms, Katya Walter's book encompasses the melding of Science and Metaphysics in which she intersperses and interweaves a personal dream with frontier science. Katya's writing skills are extensive and second to none, coining phrases that are truly inspired and unique.

Don Switlick
Institute for Neuroscience & Consciousness Studies
-:≡:-

Replete with deep scientific insights that answer previously unanswerable questions yet accessible to lay readers, Ms. Walter's book offers the most comprehensive and useful T.O.E. ever. Comprehensive in that it not only explains reality from subatomic levels to the most macro perspectives, but it also links the physical and the spiritual and connects the evolution of the universe to the evolution of consciousness. Useful in that its elegant explanation of physical reality has implications that naturally lead one to contemplate how to live one's life more effortlessly and authentically. I highly recommend this book to all those truly thirsting to understand everything.

Kevin Blackwell
Stocks & Bonds Analyst
-:≡:-

Have you ever wondered why we (human species), considering that which most concur is ineffable, continue effing it up? I've always thought having minds dead set on figuring things out in combination with phenomenon that exceed our ability to do so could be called 'God's dirty trick.' Dr. Walter has taken just such issues and playfully made a case worthy of consideration while mercifully maintaining the topic's ultimate ineffability. I found myself intellectually giggling throughout this read. Her consideration of the I Ching and our DNA alone is worth the purchase of her books.

MIL
Reader

-:≡:-

I have read several other TOE books...the latest being Tom Campbell's My Big TOE. Wanted to read this one and see what new perspectives were explored. Was pleased to find that BOTH books stand on their own, and each adds new information without contradicting the other!! (So this is a "Must Read"!)

Great visual-inducing analogies and metaphors. Enjoyed reading about her personal background, leading to development of this book.

Descriptive down-to-earth language, even tho, on occasion, I have to look up a word in the dictionary!...which means you learn new concepts and words as you go.... The subject matter covers questions we all ask at one time or another...and the answers are creative and original...makes you think and gives perspective.

Hyphenated, descriptive words are used where needed, supporting the requirement for "hyphenated sciences" and new words to explain some of the more ephemeral aspects of mind and consciousness. Very insightful, creative application of recently-discovered fractal phenomena to explain its basic principles to everything in the universe. Plenty of references and web sites for the reader who wishes to explore further.

James Beal, Ph.D.
Aerospace & Electrical Engineer

-:§:-

Katya Walter's series starting with the *Double Bubble Universe* integrates immense questions and insightful answers about the cosmos. She uses data, analogies, graphs, images, and stories that resolve together into one bubbling statement. A must read....

Rowena Pattee Kryder, Ph.D.
Dynamics & Foundations of Co-Creation

-:§:-

Katya Walter is that rare writer who can merge so-called opposing systems, like science and metaphysics. For me, with a PhD in Literature and Communication, and a serious teacher of A Course In Miracles, and having had spiritual experiences myself, I am so grateful that she brings it all together, so I no longer have to wonder if I should trust those marvelous "intuitive" experiences enough to share them with others, without fear of ridicule. Just read Katya Walter if you think this is not possible. Thanks, Katya. We need you.

Helen Bonner, Ph.D.
Author

-:§:-

Note from the Author

I FIND MANY MORE PEOPLE READ THIS BOOK THAN BOTHER TO REVIEW IT.
IF THIS BOOK WAS INTERESTING TO YOU, PLEASE REVIEW AND RATE IT.

Bibliography

This is by no means all the books I consulted on writing this series, but here are what seemed to me most relevant to this volume.

Abraham, Ralph. *Chaos, Gaia, Eros: A Chaos Pioneer Uncovers the Three Great Streams of History.* San Francisco: Harper. 1994.

Aczel, Amir D. *God's Equation: Einstein, Relativity, and the Expanding Universe.* New York: Four Walls Eight Windows. 1999.

Allgood, Kathleen T., Sauer, Tim, and Yorke, James A. *Chaos-an introduction to dynamical systems.* New York: Springer-Verlag. 1996.

Barrow, John D. & Frank Tipler. *The Anthropic Cosmological Principle.* Oxford: Oxford University Press. 1988.

Bentov, Itzhak. *Stalking the Wild Pendulum: On the Mechanics of Consciousness.* New York: Bantam Books, 1981.

Bergmann, Peter G. *The Riddle of Gravitation.* New York: Charles Scribner's Sons, 1987.

Bohm, David. *Wholeness and the Implicate Order.* London: Routledge & Kegan Paul. 1980.

…also see Myers & Briggs Foundation: /http://www.myersbriggs.org/

Campbell, Joseph. *The Masks of God.* New York: Viking Press. 1964.

Carter, Brandon. "Large Number Coincidences and the Anthropic Principle in Cosmology," in *Confrontation of Cosmological Theories with Observation.* Dordrecht: Reidel. 1974.

Casselman, Anne. "Strange but True: The Largest Organism on Earth Is a Fungus," *Scientific American.* October 4, 2007.

Casti, John L. *Alternate Realities: Mathematical Models of Nature and Man.* New York: John Wiley & Sons. 1989.

Chu, W.K. *Astrology of the I Ching.* New York: Samuel Weiser. 1976.

Chittick, William C. The Self-Disclosure of God: *Principles of Ibn al- 'Arabi's Cosmology.* Albany: State University of New York Press. 1998.

Cole, K.C. *Sympathetic Vibrations: Reflections on Physics as a Way of Life.* New York: Bantam. 1984.

Cook, Norman D. *The Brain Code.* London: Methuen. 1986.

Coveney, Peter and Highfield, Roger. *The Arrow of Time: A Voyage Through Science to Solve Time's Greatest Mystery.* New York: Fawcett Columbine. 1990.

Cowen, Ron. "Loops of Gravity: Calculating a foamy quantum space-time" in *Science News*, Vol. 153. June 13, 1998.

Culling, Louis. *The Incredible I Ching.* New York: Samuel Weiser. 1969.

D'Aquili, Eugene G. "Senses of Reality in Science and Religion: a Neuro-epistemological Perspective" in *Zygon*, Vol. 17, No. 4, December 1982.

Davies, Paul. *The Cosmic Blueprint.* London: Unwin Paperbacks. 1989.

…*Space and Time in the Modern Universe.* Cambridge: Cambridge University Press. 1977.

Davis, Philip and Hersh, Reuben. *The Mathematical Experience.* Boston: Houghton Mifflin. 1981.

DeWitt, B. and Graham, N, editors. *The Many-worlds Interpretation of Quantum Mechanics.* Princeton: Princeton University Press. 1973.

Dirac, Paul A. M. *Lectures on Quantum Mechanics.* New York: Belfer Graduate School of Science, Yeshiva University. 1964

Doczi, György. *The Power of Limits.* Boulder,; Shambhala Publications. 1981.

Dyson, Freeman. *Omni* magazine, p. 68. August, 1986.

…*Infinite in All Directions.* New York: Harper & Row. 1988.

Ellis, George F.R. and Williams, Ruth M. *Flat and Curved Space-times.* Oxford: Clarendon Press. 1988.

Eddington, Arthur. *Fundamental Theory.* London: Cambridge University Press. 1946.

Emerson, Ralph Waldo. "The Over-Soul" Essay IX from *Essays:* First Series. 1841.

Esposito, Giampiero; Marmo, Giuseppe; Sudarshan, George. *From Classical to Quantum Mechanics: An Introduction to the Formalism, Foundations and Applications.* Cambridge: University of Cambridge. 2001.

Fuller, R. Buckminster. Applewhite, E. J. *Synergetics: Explorations in the Geometry of Thinking.* New York. Macmillan. 1975.

Gardner, Martin. *The New Ambidextrous Universe: Symmetry and Asymmetry from Mirror Reflections to Superstrings.* Revised, updated edition. New York: W.H. Freeman and Company. 1990.

Gefter, Amanda. "What kind of bang was the big bang?" *New Scientist.* June 29. 2012.

Gleick, James. *Chaos: Making a New Science.* New York: Viking. 1987.

Gleiser, Marcelo. *The Dancing Universe: From Creation Myths to the Big Bang.* New York: Dutton. 1997.

Gopi Krishna. *Kundalini: The Evolutionary Energy in Man.* London: Stuart & Watkins. 1970.

Gribbin, John. *Genesis: The Origin of Man and the Universe.* New York: Delacorte Press. 1981.

...*In the Beginning: After COBE and Before the Big Bang.* Boston: Little, Brown. 1993.

Guth, Alan. *Inflation and the New Era of High-Precision Cosmology.* MIT Physics Department annual newsletter, 2002

Harrimein, Nasim. DVD: *Crossing the Event Horizon: Rise to the Equation.* Hawaii: Resonance Project Foundation. 2003.

Hobson, J. Allan, and McCarley, Robert W. "The Brain as a Dream State Generator" in *The American Journal of Psychiatry,* 134:12, December 1977.

Huajie Liu. "A Brief History of the Concept of Chaos." Peking University, Beijing, China. *http:// huajie.tripod.com/Paper/chaos.htm*

Jones, Roger S. *Physics As Metaphor.* New York; Meridian, New American Library, 1982.

Jung, C.G. *Collected Works of C.G. Jung.* Princeton: Bollingen Press. 1959.

Kaku, Michio. *Hyperspace: A Scientific Odyssey Through Parallel Universes, Time Warps and the Tenth Dimension.* New York: Anchor Books. 1995.

Li, T.Y. and Yorke, J.A. "Period Three Implies Chaos" in *The American Mathematical Monthly,* Vol. (10), 197 pp. 985-992. 1975.

Lindley, David. *The End of Physics: The Myth of a Unified Theory.* New York: Basic Books. 1993.

Lisi, A. Garrett. *An Exceptionally Simple Theory of Everything:* http://arxiv.org/abs/0711.0770

Lorenz, Edward. "Predictability: Does the Flap of a Butterfly's Wings in Brazil Set Off a Tornado in Texas?" *Washington: American Association for the Advancement of Science.* December 29, 1979.

...*The Essence of Chaos.* Washington: University of Washington Press. 1993.

McFadden, Johnjoe. *Quantum Evolution: How Physics' Weirdest Theory Explains Life's Biggest Mystery.* New York: W.W. Norton & Company. 2001.

Morris, Richard. *Dismantling the Universe: The Nature of Scientific Discovery.* New York: Simon and Schuster. 1983.

...*The Universe, the Eleventh Dimension, and Everything: What we Know and How We Know It.* New York: Four Walls Eight Windows. 1999.

Noble, David F. *The Religion of Technology: The Divinity of Man and the Spirit of Invention.* New York: Alfred A. Knopf. 1997.

Pagels, Heinz. *The Cosmic Code: Quantum Physics as the Language of Nature.* New York: Simon and Schuster. 1982.

Penrose, Roger. *The Emperor's New Mind.* Oxford: Oxford University Press. 1989.

Pickover, Clifford A. *Black Holes: a Traveler's Guide.* New York: John Wiley & Sons, Inc. 1996.

Pomeranz, Kenneth. *The Great Divergence: China, Europe, and the Making of the Modern World Economy.* Princeton University Press, 2000.

Pratyagatmananda, S. The Metaphysics of Physics. Madras, India: Ganesh. 1964.

Prigogine, Ilya. Order Out of Chaos: Man's New Dialogue With Nature. With Isabelle Stengers. New York: Bantam Books. 1984.

Prkic, Lada. "Geometry, All Around Us" at https://www.linkedin.com/pulse/geometry-all-around-us-lada-prkic-ceng/

Puri, Lekh Raj. *Mysticism-The Spiritual Path.* New Dehli: Radha Soami Satsang Beas.

Raymo, Chet. *Skeptics and True Believers: The Exhilarating Connection Between Science and Religion.* New York: Walker and Company. 1998.

Roszak, Theodore. *The Gendered Atom: Reflections on the Sexual Psychology of Science.* Berkeley: Conari Press. 1999.

Russell, Bertrand. *Introduction to Mathematical Philosophy.* London: Allen & Unwin. 1919.

Sarfatti, Jack. Psychoenergetic Systems, "Implications of Meta-Physics for Psychoenergetic Systems" in Vol. 1. London: Gordon and Breach. 1974.

...and Toben, Bob. *Space-Time and Beyond.* New York: Harper and Row. 1968.

Schonberger, Martin. *The I Ching and the Genetic Code: the Hidden Key to Life.* New York: ASI Publishers. 1979.

Sessions, Roger. *Wisdom's Way: The Christian I Ching.* The Christian I Ching Society of Cedar Park, Texas. 2015.

Sheldrake, Rupert; McKenna, Terence; Abraham, Ralph; Houston, Jean. *Chaos, Creativity, and Cosmic Consciousness.* Rochester, Vermont: Park Street Press. 1992

Shubnikov, A.V., and Koptsik, V.A.; trans. editor, Harker, David. *Symmetry in Science and Art.* New York: Plenum Press, 1974.

Susskind, Leonard. "Black Holes & the Information Paradox" in *Scientific American,* April, 1997.

Susskind, Leonard; Lindesay, James. *An introduction to black holes, information and the string theory revolution: The holographic universe.* River Edge, New Jersey: World Scientific Publishing Company, 2004.

Tiller, William. *Conscious Acts of Creation: the emergence of a new physics.* Walnut Creek, California: Pavior Publishing. 2001.

...*Science and Human Transformation: Subtle Energies, Intentionality, and Consciousness.* Walnut Creek, California: Pavior Publishing. 1997.

Vance, Murl. *The Trail of the Serpent.* Poona, India: Oriental Watchman Publishing House.

Von Franz, Marie-Louise. *Number & Time.* Evanston: Northwestern University Press. 1974.

Weinberg, Steven. *Gravitation and Cosmology: Principles and Applications of the General Theory of Relativity.* New York: John Wiley and Sons, 1972.

...*The First Three Minutes (updated): A Modern View of the Origin of the Universe.* New York: Basic Books, 1988.

Weyl, Hermann. The Classical Groups, their Invariants and Representations. Princeton University Press, 1939. See p. 165.

Whaley, Arthur. *Three Ways of Thought in Ancient China.* New York: Doubleday Anchor Books. 1939.

Wheeler, John A. *Geons, Black Holes and Quantum Foam: A Life in Physics.* New York: Norton and Company. 1998.

Wigner, Eugene. *Symmetries and Reflections.* Westport, Conn.: Greenwood Press. 1967.

...Invariance in Physical Theory in American Philosophical Society Proceedings, Vol. 93, No. 7.

Wilhelm, Richard (Translator); Baynes, Cary F. (Translator); Wilhelm, Hellmut (Preface); C. G. Jung (Foreword). *I Ching or Book of Changes.* Princeton NJ: Princeton University Press, 3rd. ed., Bollingen Series XIX, 1967; 1st ed. 1950.

www.ingramcontent.com/pod-product-compliance
Lightning Source LLC
Chambersburg PA
CBHW060527210326
41519CB00014B/3148